FIASCO AT 1280

THE RISE AND HARD FALL OF A TWIN CITIES RADIO STATION

JEFF R. LONTO

STUDIO Z-7 PUBLISHING
MINNEAPOLIS, MN.

NOTICE: This book was not created, authorized or otherwise endorsed by any current or former owners of radio station WWTC.

International Standard Book Number: 0-9660213-4-7

Published by Studio Z-7 Publishing
813 Marshall Street NE
Minneapolis, MN 55413-1816

Cover Design: Jeff R. Lonto

Printed in the USA by

3212 East Highway 30 • Kearney, NE 68847 • 1-800-650-7888

For Sandy Charles

On behalf of her friends who miss her.

(Photo Courtesy Larry Wolf)

ACKNOWLEDGMENTS

When I took on this project a number of years ago, I had no idea of the kind of enthusiastic support I would get from the people who were directly involved with radio station WWTC. They have been tremendously generous with their time, resources and friendship.

A very special thanks and a tip of the hat to: Adam Abrams, Roger Bull, Scott Carpenter, Dick Driscoll, Ray Erick, John Farrell, Al Flom, Arne Fogel, Styx Franklin, Stuart Held, Michael Jaye, Arthur Phillips, Brad Piras, Darrel Renstrom, Del Roberts, Nancy Rosen, Mike Ryan, Scott Stevens, Brian Turner and Larry Wolf. This book would not have been possible without their input.

I would also like to thank *Star Tribune*, the *St. Paul Pioneer Press*, David Carr of the Washington, D.C. *City Paper*, Tom Bartel of *City Pages*, the Minneapolis Public Library, Dennis Rose, Ed Brouder and Man From Mars Productions and Ms. Marcia Reich, without whom the book, nor the author, would be possible.

Finally, to everyone who supplied me with the equipment and the facilities for this project to materialize, and to everyone who gave me encouragement through my frequent bouts of frustration and so-called "writers' block", may I please offer my deepest gratitude.

CONTENTS

INTRODUCTION

The AM radio dial today isn't what it used to be. Although the recent resurgence in talk radio has prevented the death of the frequency band that started it all, the age of FM stereo and digital communications dominate the listening habits of most people.

It is often hard to believe but it wasn't very long ago that the radio stations most of us listened to were on the noisy, static-filled AM dial. While FM, which has been around since the 1940s, was reserved primarily for classical music and "long-hair" programming, AM had the news, sports, rock, country and popular music. Only the well-to-do had FM radio in their cars and the portable transistor radios we held to our ears as kids were certainly not FM equipped. There wasn't anything in that spectrum we were interested in hearing anyhow.

Our parents listened to the AM "middle-of-the-road" stations which played mostly soft popular music and standards with a lot of news and sometimes play-by-play sports. Stations such as KDKA Pittsburgh, WGN Chicago, KGO San Francisco and WCCO in the Twin Cities had phenomenal success with that format.

Teenagers and young adults often made top-40 rock 'n' roll stations even more successful. Stations such as WABC New York, WLS Chicago, KHJ Los Angeles and two stations, WDGY and KDWB in the Twin Cities, burned up the ratings. With a tight playlist of music, fast-talking personality-oriented disc jockeys, call-letter jingles and lots of contests, the AM band in almost every city had its own youth haven. It helped define what came to be known to some as the culture of "youth rebellion", although it was financed by the same corporations that sold products on and owned the radio stations our "establishment" parents listened to.

It seemed rock 'n' roll radio belonged to the young people of America. Those who grew up on AM rock radio tuned in not only for the music but for their favorite disc jockey as well. It was a time when the DJ did more than play records, announce call-letters and read weather reports; the jocks were talented personalities that gave a little of themselves to us when they were on the air.

They made us feel good by playing our favorite songs and making us laugh at the same time. Girls had crushes on them and boys thought of them as part of the gang. Even the lonely kid had a friend on the radio who would always be there. We could talk to them on the phone when we called the request line and meet them and get their autograph when they made public appearances.

By the late 1970s, as audiences became more sophisticated, FM gained momentum in the arena of popular music by promising not only a higher quality signal, but "more music with less talk". While some radio buffs and purists may argue that FM embodies a heartless, soulless, detached, cold technology, AM stations were nonetheless rapidly abandoned in favor of FM, especially as more and more automobiles were FM equipped.

AM radio stations, especially those without FM sister stations, struggled for dear life in the late seventies and into the eighties. Stations with old-line disc jockey formats had a particularly hard struggle.

Although in the Minneapolis-St. Paul metropolitan area, WCCO-AM, the 50,000-watt clear-channel "Good Neighbor to the Northwest" continued to dominate even the FM stations for years to come (although it gradually lost ground and eventually fell to second place in favor of KQRS-FM), the big top-40 stations had to struggle with the changing times.

KDWB began to shift its focus toward its own newly-acquired FM station in the late '70s and WDGY, defeated by the competition it inspired as the first rock station in Minnesota, switched to country music. Other AM stations began to rediscover the two-way talk format. In the age of high-fidelity stereo sound, rock (and other music formats) was dead on AM radio. The fast-talking, high-personality disc jockeys went with it, replaced by cold, smooth, silky-voiced announcers.

And then there was WWTC.

WWTC was the enigma of Twin Cities radio. An old, struggling station with no FM counterpart, WWTC did not become one of the great AM rock stations until, ironically, after the AM rock radio went out of vogue.

WWTC had a reputation for being one of the more unique stations in town. In the late 1970s, it was one of the few news-talk stations in the country. When that format failed, the station began hiring some sixties-era rock jocks and others who worked together and built their own ideal radio station.

WWTC entered the 1980s in a time warp. The station not only played the hits of the '50s, '60s and '70s when nobody else was, but the disc jockeys were once again giving a little of themselves to us. Once again, we could talk to them on the phone when we called the request line and we could meet them and get their autograph. The WWTC disc jockeys made us laugh and forced us to listen on more than a passive level.

The WWTC call-letters had their own history and mystique in the Twin Cities area which added to the station's nostalgic charm.

Unfortunately, changing trends, low budgets, bad decisions and other factors limited WWTC's potential and the format which that station is best remembered for to this day, lasted only half a decade. After doing everything from golden oldies, news-talk and middle-of-the-road music to urban dance, all-weather and fusion jazz, WWTC seemed to find its niche as "Radio AAHS", catering to listeners who are predominantly too young to fill out a ratings diary. But even that hit the pot holes in the mean streets of the business.

Today, the radio dial looks a lot like cable television; stations are much more into specialized programming rather than doing a little bit of everything. Radio is becoming more defined all the time, with "niche" formats. Listeners are thought of less as people and more as "target demographics".

Stations today do only news or play only hard rock or country or funk or jazz with no room for crossover. Even the hard rock category in recent years has broken down into formats such as classic rock, modern rock and heavy metal.

The story of WWTC is one of the great survival stories in Twin Cities broadcasting. It has been known to succeed in spite of itself and fail because of itself, yet has managed, amazingly, to stay afloat.

WWTC, with all its fowl-ups, bleeps and blunders over the years, is the strange creature that just wouldn't die.

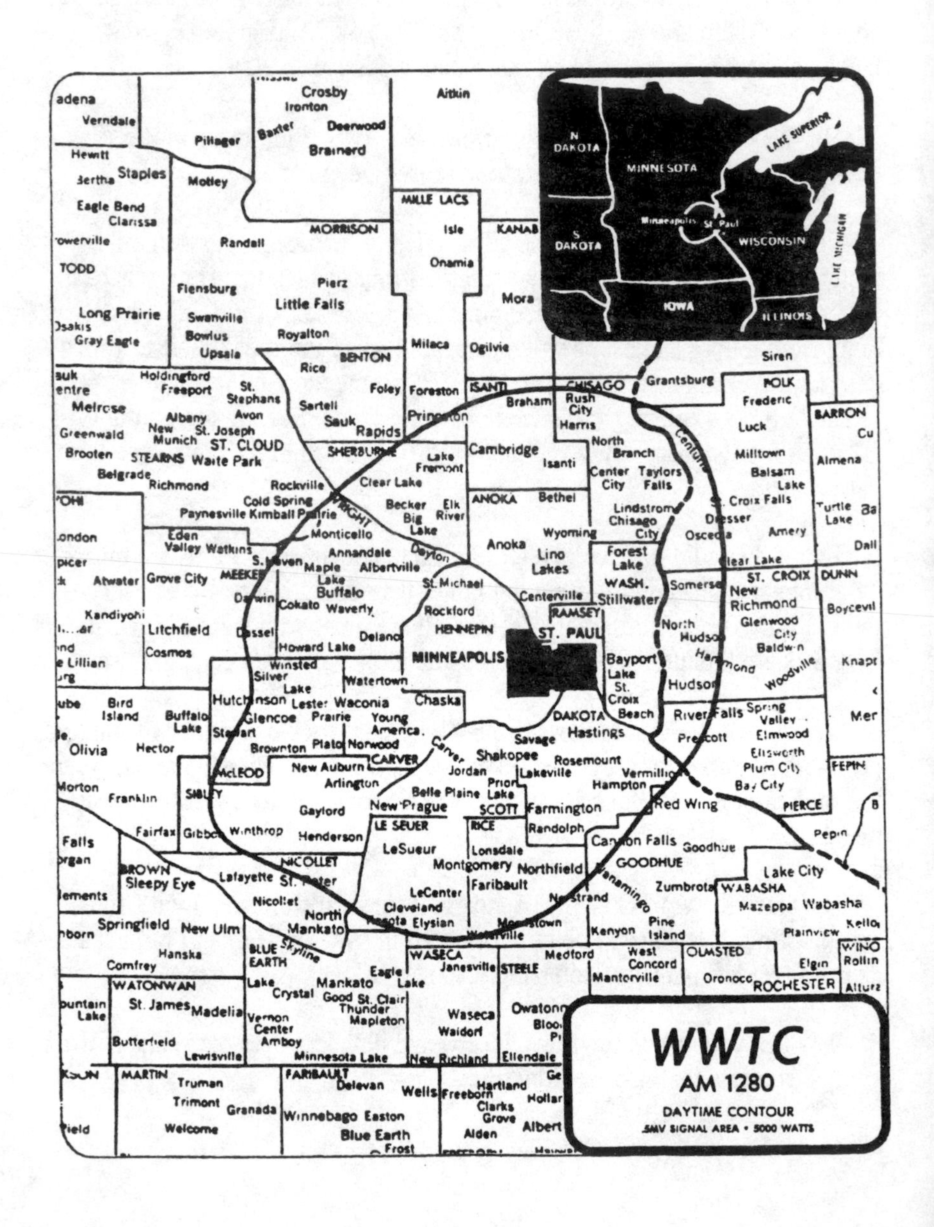
N DAKOTA
MINNESOTA
LAKE SUPERIOR
S DAKOTA
Minneapolis
St. Paul
WISCONSIN
LAKE MICHIGAN
IOWA
ILLINOIS
Crosby
Ironton
Aitkin
Verndale
Pillager
Baxter
Deerwood
Brainerd
Hewitt
Staples
Motley
Eagle Bend
Clarissa
MILLE LACS
MORRISON
Isle
KANAB
Randall
Onamia
TODD
Flensburg
Pierz
Little Falls
Mora
Long Prairie
Swanville
Bowlus
Royalton
Gray Eagle
Upsala
Milaca
Ogilvie
BENTON
Siren
Holdingford
Freeport
St. Stephans
Avon
Rice
Foley
Foreston
ISANTI
CHISAGO
Grantsburg
POLK
Melrose
Albany
Sartell
Braham
Rush City
Frederic
Greenwald
New Munich
St. Joseph
Sauk Rapids
Princeton
Harris
Luck
BARRON
Brooten
STEARNS
ST. CLOUD
Waite Park
SHERBURNE
Lake Fremont
Cambridge
North Branch
Isanti
Milltown
Almena
Belgrade
Richmond
Rockville
Clear Lake
Center City
Taylors Falls
Balsam Lake
Cold Spring
Paynesville
Kimball
Prairie
Becker
Big Lake
Elk River
ANOKA
Bethel
Lindstrom
Chisago City
St. Croix Falls
Dresser
Osceola
Amery
Turtle Lake
WRIGHT
Eden Valley
Watkins
Monticello
Dayton
Anoka
Wyoming
Lino Lakes
Forest Lake
Clear Lake
London
Annandale
Albertville
St. Michael
S. Haven
Maple Lake
Buffalo
Atwater
Grove City
MEEKER
Darwin
Cokato
Waverly
Rockford
Centerville
WASH.
Stillwater
Somerset
ST. CROIX
New Richmond
DUNN
Glenwood City
Boyceville
Kandiyohi
Litchfield
Dassel
HENNEPIN
RAMSEY
ST. PAUL
North Hudson
Baldwin
Cosmos
Howard Lake
Delano
MINNEAPOLIS
Bayport
Hammond
Woodville
Winsted
Silver Lake
Watertown
Lake St. Croix Beach
Hudson
Bird Island
Hutchinson
Lester Prairie
Waconia
Chaska
DAKOTA
River Falls
Spring Valley
Buffalo Lake
Glencoe
Young America
Hastings
Prescott
Elmwood
Olivia
Hector
Stewart
Brownton
Plato
Norwood
Savage
Ellsworth
Plum City
CARVER
Carver
Shakopee
Rosemount
McLEOD
New Auburn
Jordan
Prior Lake
Lakeville
Vermillion
Hampton
Bay City
Morton
Franklin
SIBLEY
Arlington
Belle Plaine
Red Wing
PIERCE
Gaylord
New Prague
SCOTT
Farmington
Fairfax
Gibbon
Winthrop
Henderson
LE SEUER
RICE
Randolph
Cannon Falls
Goodhue
Pepin
LeSueur
Lonsdale
GOODHUE
NICOLLET
BROWN
Sleepy Eye
Lafayette
St. Peter
Montgomery
Northfield
Faribault
Wanamingo
Lake City
Zumbrota
WABASHA
Nicollet
LeCenter
Nerstrand
Mazeppa
Wabasha
Springfield
New Ulm
North Mankato
Cleveland
Elysian
Morristown
Waterville
Kenyon
Pine Island
Plainview
Skyline
Hanska
BLUE EARTH
WASECA
Janesville
STEELE
Medford
West Concord
OLMSTED
Elgin
Rollin
Comfrey
WATONWAN
Eagle Lake
Mantorville
Oronoco
ROCHESTER
Lake
Crystal
Mankato
Good Thunder
St. Clair
Mapleton
St. James
Madelia
Vernon Center
Amboy
Waseca
Waldorf
Butterfield
Lewisville
Minnesota Lake
New Richland
Ellendale
MARTIN
FARIBAULT
Truman
Delevan
Wells
Freeborn
Hartland
Trimont
Granada
Winnebago
Easton
Clarks Grove
Welcome
Blue Earth
Alden
Albert
Frost
WWTC
AM 1280
DAYTIME CONTOUR
.5MV SIGNAL AREA • 5000 WATTS

1. THE LAST OF THE AM ROCKERS

Fear and Loathing on the AM dial

Scott Carpenter was a recently divorced 35-year-old native Minnesotan. A graduate of Brown Institute, the famed broadcast school in Minneapolis, he had acquired a strong reputation in his 15 years as a broadcaster, having worked in major cities all over the country.

When Carpenter returned to his home turf to save a dying Twin Cities radio station, he quickly realized he had walked into a mine field.

WWTC, a 5,000-watt AM station at 1280 on the dial was in shambles in the summer of 1979, when Carpenter arrived. WWTC was playing "adult contemporary" music at the time, having recently dropped a news-talk format due to low ratings. The current format was doing even worse. The station had been losing money for years and morale around the place was volatile.

The air staff consisted almost entirely of reporters and anchors from the old news format, including, John Farrell, Doug McLeod, Dave Hellerman and Deanna Burges, as well as engineer Al Flom and a number of sales people and others. Most of the announcing staff knew little about music radio and were uncomfortable as disc jockeys. Resentment toward the management and especially the owners was intense. The new program director was viewed with as least as much disdain.

Carpenter had been hired through the national consulting firm of Burns Media. Radio personalities tend to have contempt and disdain for hired consultants, outsiders who come to a station usually from another city, change everything around based on computer analysis and research with no regard whatsoever to the local community and flavor. People are hired and fired based on the decisions of consultants. Carpenter not only represented the lowly consultants, but an even bigger foe as well. "I was the new guy coming in after the news people had been blown out and there was a tremendous amount of animosity", Carpenter recalls, "I had never been there before but I was an absolutely hated man because I represented *them.* I represented the enemy."

The "enemy" was Robert Short, an infamous local businessman who had purchased WWTC Radio in 1978, with his seven children in partnership, during his unsuccessful campaign for U.S. Senate. The station was still programming news and talk at the time.

Short and the WWTC staff had a mutual hatred for each other that went back almost to the day he bought the station. When Short chastised the staff for the station's poor financial situation and told news reporters they would be expected to help the sales department sell advertising, the news people responded by unionizing. During Short's Senate campaign, the station ran editorials against him. Once Short called a staff meeting and fired several people in front of everyone else.

The general manager, a soft-spoken older gentleman by the name of Wayne (Red) Williams, had leadership qualities that many at the station found questionable.

"Red Williams was incompetent", opines Al Flom, the station engineer. "He was nice old guy but he was incompetent. He'd been hit in the head too many times playing football in college."

Indeed Williams, who had been hired just a few months earlier, had managed WLOL AM-FM in St. Paul throughout the sixties and seventies but those stations, with all the different formats that were tried, were never ratings leaders during his long tenure there.

Carpenter realized he had a serious morale problem. The ratings were bad, the owners were threatening to shut the place down or sell it to Minnesota Public Radio, the pay was at or near minimum wage, fights would literally break out in meetings between management and staff and the sales staff didn't make much a priority of selling advertising. There was bad blood between the staff and management, and even worse blood with the ownership.

Turning around WWTC would be Carpenter's biggest challenge yet; there wasn't much that could be programmed on an old-line AM station that wasn't already being done better on FM. AM radio was on the decline and declared all but dead in the late 1970s. He would also be expected to work on a shoe-string budget, as the ownership was extremely cost-conscious.

Carpenter slowly gained the trust of the employees and even helped those who didn't want to stay on board find new jobs elsewhere. He started interviewing new people and took a hard look at what the station was programming.

It was clear the current format of soft pop hits couldn't continue. According to the ever-important Arbitron ratings, which dictate the number of people tuned in and thus, the amount of money that can be charged to advertisers, the format earned WWTC just six-tenths of one percent of the local listening audience, down from 1.7 percent in its last ratings period as a news station. It was the lowest share in the station's history.

Although the old and respected WCCO-AM (830) still dominated the ratings, almost all the top-rated stations were on the static-free FM dial. It just wasn't hip to listen to AM radio anymore. Carpenter sought to change that.

It was decided that the only way to make a faltering AM station successful is to do something nobody else is doing and do it well. It would have to be something unique that an FM station wouldn't likely touch anytime soon. It would have to be something that people would actually enjoy hearing on low-fidelity radio.

Carpenter began to experiment with rock 'n' roll oldies. The "adult contemporary" format was already including some not-so-contemporary music, mostly from the 1960s. Carpenter began to program "solid gold" from the 1950s twice an hour.

Because the music of the fifties and early sixties was recorded in mono, and the advent of digital technology was yet to come, old time rock 'n' roll was the one format that would sound good on an AM station. AM radio was where most children of rock 'n' roll first heard their favorite songs so WWTC's low-fidelity signal would actually create an ambience of nostalgia that would add to the station's appeal.

With the addition of Buddy Holly, the Platters and Little Richard to the station's music selection, Carpenter began searching for "fun" disc jockeys, people who knew the art of being master of ceremonies rather than just announcers who knew only how to read slogans and weather reports. As the old news-turned-disc jockey staff began finding new gigs elsewhere, Carpenter put out word that he was going to build an all-new WWTC.

Birth Of a Station

Scott Carpenter was a something of an eccentric. Most who had worked with him recall him as intelligent, creative and somewhat temperamental. He was also controversial; people either liked him a lot or they couldn't stand him. He had a specific concept of how a radio station should be run and a bit of an attitude.

In addition to his duties as program director, he began hosting the morning show on WWTC. He told jokes, he had in-studio guests and he made his show interactive by encouraging listeners to call the studio and go on the air with him for a contest or comedy skit. His on-air signature was "You got Carpenter in your radio!"

The first disc jockey he brought to WWTC was "Doctor" Dave Gross, who had been working in Duluth. Carpenter literally discovered him while vacationing there. Dr. Dave was a wild, screaming and yelling kind of rock 'n' roll DJ — something that had not existed in the tame atmosphere of Twin Cities radio in a long time. Amazingly, he was recruited from the far more conservative Iron Range.

Also hired was Del Roberts, an old friend he had worked with in the late sixties to do weekend and occasional fill-in shifts at the station. Roberts called himself "Ugly Del" on the air and spiced his show with a sardonic, witty sense of humor.

Meanwhile, Brad Piras, a part-time announcer on a small FM jazz station in town, had heard WWTC was hiring. He called Carpenter on the request line while Scott was doing his show.

"Why should I see you?" Carpenter asked while a song played.

"Well, because I'm good. I can make good things happen."

"Yeah? I get a lot of calls from people who say that."

Carpenter was actually beginning the job interview right there, trying to prod the young disc jockey to "wow" him. He finally did ask Piras to come down for a formal interview and it would be one that Piras wouldn't soon forget.

A firm believer in the power of positive thinking, Carpenter would give motivational speeches to the staff with a mix of humor and philosophy. He would call the disc jockeys while they were on the air and critique their "bits", always making a point of telling them that they're doing a good job as well as criticizing.

On Friday nights, he would sometimes bring the crew to the Court Bar and Grill, located down the block from the WWTC studio in downtown Minneapolis. The Court was a popular hangout for radio personalities as the WCCO studio was around the corner.

WWTC became known as the home of "The Twin Cities Greatest Hits". A request line was set up and as listeners were hearing and requesting Chuck Berry, Elvis, Herman's Hermits and the Monkees on the radio for the first time in years, in addition to the worn out hits everyone else was playing. The response was tremendous. There were indications that WWTC was on to something.

Since the station had to make itself known on an incredibly small budget, Carpenter looked at ways to gain free publicity through stunts and other off-the-wall promotions. He was known to make prank phone calls on the air and sometimes go out in public to pull some kind of gag to make WWTC visible. The call-letters were still synonymous with news radio to most people so Carpenter's task was to give WWTC a "fun" reputation.

In September, while the weather was still warm, he moved his office to the roof of the downtown Minneapolis building where WWTC was located. Passers by and people working in offices adjacent from the building could see him up there, sitting at a desk with a typewriter and telephone, supposedly working. (It was fairly simple to move everything up there because of stairs inside the building that lead to the roof.)

"I'd been saying something [on the air] about how I gotta do something about that office that's right next to the general manager's", Carpenter recalls, "and there's no question about it, I was having problems with the general manager at the time because he wouldn't leave me alone! It was a really bad office. It was more like a hallway. So I'm talking about it on the radio and one morning a guy calls up and said 'just put your office on the roof'".

Scott and the other jocks began going up there regularly, waving to the secretaries and others in neighboring buildings. "Doctor" Dave Gross even went as far as bringing flowers to the secretaries, pointing to Scott through the window, who would wave.

General manager Red Williams had a problem with this seemingly silly publicity stunt but Bob Short, the stern, cantankerous owner of WWTC surprisingly approved. "For Christ's sake do it!" Carpenter recalls being told by Mr. Short.

Something's Burning

In the fall of 1979, the beleaguered AM radio station was finally going somewhere. The format was becoming more cohesive, the air personalities seemed to be enjoying themselves and listeners and sponsors were beginning to take notice. WWTC had nowhere to go but up and it seemed to finally be doing that.

But if anything can possibly go wrong, it will. On September 23, 1979, a Sunday afternoon while weekend disc jockey "Ugly" Del Roberts was on the air, listeners heard a few pops and cracks and then — nothing; the transmitters, located at a remote site about five miles away from the studio, were on fire.

"That was caused by somebody turning on the auxiliary transmitter into a dummy load", recalls engineer Al Flom. "It was on the wall, it was just for testing purposes, only meant to be left on for fifteen minutes. It was left on for hours. It heated up the resistors that were in the wall and burned up the transmitter building."

Flom was called on to get the station back up and running. Flom, who was the station's chief engineer for almost ten years, had actually quit WWTC two weeks earlier. But since he knew the transmitters better than anyone else, he was hired back on a free-lance basis.

He strung together parts of the two damaged transmitters and got the station back on the air two days later with about 80 percent power. The old muddy-sounding RCA transmitters were outdated and had been in use for some 25 or 30 years, back when the station was known as WTCN.

The old, rebuilt transmitter was unstable, however, and had to be monitored on a 24-hour basis until a new one could be installed. The station hired some people to babysit it, watching the meters and taking hourly readings.

Flom, meanwhile, went to work for a production company in Minneapolis called Voiceworks, although he continued to do free lance work for WWTC.

"The reason I left", Flom says, "was because I got a call from NSP [Northern States Power] that said 'If you don't pay your bill by three o'clock this afternoon we're shutting your power off', which for a radio station would be real cool". It was the responsibility of the owner, not the engineer to keep the electric bill paid.

"Short ran that thing on a cash-only basis for everything that got delivered," Flom says. "He just wouldn't pay until they took him to court and he paid maybe ten cents on the dollar, so everything had to be C.O.D."

Mike McKenzie replaced Al Flom as chief engineer at WWTC. Ironically, Flom had been McKenzie's assistant engineer there in the early 1970s until McKenzie left a few years later, making Flom chief engineer.

Meanwhile, Brad Piras, the young disc jockey who had been "interviewed" over the telephone by the program director, came down for his more official interview. He drove downtown to the studio and arrived at ten a.m., went to the front desk and asked to see Scott Carpenter.

"Scott's on the roof", the receptionist told him. She gave him instructions on how to get up there. He walked up the stairs and found Carpenter in his rooftop office.

"Come on over here, sit down", Piras remembers being told. "You got a resume?"

"Yeah."

"I don't want to see it! Are you any good?"

They began to talk. Scott fixed him a drink and finally said "I tell you

Brad Piras (Courtesy Larry Wolf)

what, you're hired. You start tomorrow morning." Piras thought he was getting the morning disc jockey shift but instead, Carpenter gave him directions to the transmitter site and told him to report there for his new job of reading meters. Before long, however, Piras was in the studio doing late night local news updates and weekend DJ shifts.

The Shorts collected the insurance from the fire and purchased for the station a state-of-the-art transmitter along with an "Optimod" audio processor, The "Optimod" was a filter device that made the signal louder and crisper, thus making the 5,000-watt WWTC actually sound stronger in the metro area than the big 50,000-watt stations. The signal literally jumped out when one tuned past it on the dial.

With all the sweeping changes, it was decided a new general manager was needed, someone more open and more progressive than Red Williams.

"He would fight everything that was happening", Carpenter says. "Bob Short knew it, Burns Media was having trouble with him and I was having problems because any time I wanted to make a change he'd fight it. . .I realized as we all did that Red really needs to go away and retire."

The position was offered to Carpenter but he didn't want to take on that kind of responsibility. "In the interim until the next general manager came in, more or less I ran it with Bob Short, Brian Short, the gang. It sort of ran itself." Eventually sales manager Dennis O'Leary was named "acting" general manager until a replacement could be found.

A new, talented, youthful disc jockey staff was beginning to solidify. In addition to Dr. Dave, "Ugly" Del and Brad Piras, names such as Mike "Records" Ryan, Nancy Rosen, Steve "Boogie" Bowman and others were heard on Radio 1280, each with their own distinct personality

Carpenter continued to work through Burns Media, the consulting firm that had been hired to turn WWTC into a major profit-making operation. The "Twin Cities Greatest Hits" format was working, but research was showing that the station needed a better identity and direction if it was going to make any inroads in the nasty competition for an audience.

Bernie Torres, one of Carpenter's Burns Media colleges, did research using focus groups and studying the Minneapolis-St. Paul market. In his research he found that two words, "gold" and "rock", had the most recall impact with sample audiences.

"Bernie calls me on the phone and says 'Scott, we got it! It's Golden Rock!'", recalls Carpenter. "He said 'Do you feel good about it?' I said 'Man, I'm telling you, this is it!"

The Golden Rock

On Monday, October 1, 1979, the experimental phase was over. With a whole new sound, WWTC premiered as "The Golden Rock of Minneapolis-St. Paul". With the new transmitting equipment and the new format of golden oldies, the rinky-dink station at 1280 on the dial began to stick out like a sore thumb; one couldn't tune the AM dial and avoid it. The sound was unique; in the age of "more music, less talk", WWTC had an attitude of hype that bounced with the music it played.

"Here we are on the AM band; why don't we duplicate to some extent what it sounded like when AM was in its prime?" Carpenter rhetorically asked *Minneapolis Tribune* media critic Neal Gendler. "Songs can bring back a whole picture of a time in your life."

Along with the old music, WWTC brought back the rapid-fire, fast-talking disc jockeys that permeated AM radio when these particular records were coming out, complete with the talkovers and constant time and temperature readings.

"You got it!" a DJ would bellow out. "For the Class of '61, Curtis Lee, 'Pretty Little Angel Eyes' on the Golden Rock WWTC. What a snappy ditty! Reminds me of the time I had my ditty snapped! Three-fourteen now in the Twin Cities with 85 degrees on the outside. The humidity rate, 76 percent. The stupidity rate, 100 percent. This next tune goes out to Jack, John, Rich, Julie, Laurie and to Skinless Frank on the west side. It's the Raspberries with a little raunchy radio from 1972. 'Go All the Way' on TEE-CEE!"

It was something of a shock to the passive listener. At a time of mediocrity and blandness in almost all aspects of popular culture, at a time when it was fashionable to play a number of songs in a row with no talk, at a time when radio announcers (especially on the trendy FM stations) tried to be as sedate and inoffensive as possible both in the way they spoke and in the music they played, WWTC was in your face. It was intense and aggressive.

Although WWTC was not the first station locally to play golden oldies, the format was out of chic in the late seventies and early eighties, a time of transition not only between Jimmy Carter and Ronald Reagan, but between disco and so-called "new wave" rock. In addition, soft pop artists such as Air Supply and the Alan Parsons Project dominated radio station playlists. "Classic rock" and "oldies FM" stations would not become a fixture on the radio dial for years to come.

WWTC played the music that was assumed everyone had outgrown and had disc jockeys who seemed to do whatever they wanted. They were constantly present, talking over the beginnings of songs, joking around, giving trivia about a song or artist and putting telephone calls on the air and asking people to do strange things like slurp coffee or sing along with the next tune. They demanded attention from the listener and insisted on having fun.

The Golden Rock was interactive. Audience participation was an important part of the format and something almost magical began to develop between the listeners and the radio station itself. Phone calls were often put on the air and people would flood the studio lines to dedicate a song to a sweetheart or best friend, or reminisce what they were doing when a particular record was popular. Sometimes they would just talk about what they were doing at that moment.

As for the music, the station would play roughly two hits from the fifties per hour, four from the sixties, three from the seventies and two currents (dubbed as "future gold"), with time for commercials and DJ "bits" in between.

The party would stop for five minutes at the top of every hour as the Mutual radio network cut in for hourly news. The network also cut in at 1:35 every afternoon for a sports update and fed a daily entertainment feature, "Eye on Hollywood", to WWTC.

News actually punctuated the format. In addition to the hourly network feeds, there was local news on the half-hour during the morning and afternoon drive times anchored by Elliot "Skip" Smith. The local news on WWTC tended to emphasize the violent and sensational, with as many reports about shootings, traffic fatalities and armed robberies that could be gathered from the wire services.

Also in the mix was the classic "Chickenman" radio serial. "Chickenman", billed as "the most fantastic crime fighter the world has ever known", fought crime in the fictional Midland City wearing a white, feathered suit. Helping out were the Police Commissioner, his secretary Miss Hellfinger and Mildred, the Masked Mother.

"Chickenman" is something of a forgotten classic of AM rock radio. The silly but often amusing serial ran in five-minute daily installments (running three times daily on WWTC) that were created and produced in the 1960s by Chicago radio personality Dick Orkin and originally syndicated primarily to top-40 stations nationwide.

The new format on WWTC had plenty of skeptics. Nobody was playing oldies in the Twin Cities or almost anywhere else at the time and the whole concept seemed particularly absurd. In an age of disco, polyester and pseudo-sophistication, and with a brand new decade almost under way, who would want to hear that old music, not to mention those annoying, hyper DJs and an inane radio serial about a chicken man that had long since disappeared? It was assumed people want to hear more music and less talk. That was the mantra of just about every radio station in the country.

Competing for the scraps left over in the shadow of almighty WCCO-AM, there were several adult contemporary, country, top-40 and album rock stations on AM and FM, all of which sounded strikingly similar to each other. Contemporary radio strived for mediocrity. The theory was to not make waves, don't be creative as that will just turn people off.

WCCO, ironically, still commanded over 30 percent of the local listening audience despite its chatty format and its exclusive position on the supposedly obsolete AM dial. It was proof that an AM station can survive quite well with innovative programming. WWTC had roughly one percent of local radio listeners. It had nothing to lose by being different.

Deacons Of the Discs

The secret to WCCO's success was the well-established, well-known, likable personalities. It was fun personalities that had also made the rock stations big in decades past, before the post-war generation went to college, turned on to album cuts and became "sophisticated".

Taking a gamble that maybe this same generation was ready to lighten up and have fun again, WWTC captured the spirit of old time rock 'n' roll radio with its own lineup of personalities. Each personality established himself with his own unique style and a lot of self promotion.

Bright and early at 6:05 in the morning, immediately following five minutes of Mutual network news, the program director kicked off the day.

"Ladies and gentlemen, Scott Carpenter — and the Golden Rock of Minneapolis-St. Paul — WWTC!" an announcer bellowed out over a trumpet fanfare. Carpenter would immediately go into a rousing rock classic such as

"Shake, Rattle and Roll" or "Tutti-Frutti".

Scott was the quintessential morning personality. He talked to his audience, sympathizing with them having to get up and get ready for work and stayed with them on the drive in, starting off their day with a few laughs, upbeat music, points of interest, and news, weather and traffic updates.

When Carpenter started on WWTC, he did his show solo but he soon acquired a partner, a feminine voice to do news, be a sidekick and serve as a counter-balance to his general weirdness. Her name was Nancy Rosen.

Nancy was a petite, wide-eyed 21-year-old who had been hired from a station in San Francisco. She was a Minnesota native, as was Carpenter, and had been on the air locally at hard rock station KQRS-FM (92.5) as one of the breathy-voiced "token" female announcers in the late seventies. She had a distinct, sultry voice that served as a welcome contrast to the predominantly male-orrientated format. After Scott's shift ended at nine a.m., Nancy continued with her own show which lasted until noon.

Afternoons on WWTC were initially handled by Carpenter's protege "Doctor" Dave Gross, but unfortunately, an on-air incident lead to his dismissal.

"Dave was forced to leave and that was a damn sad thing" Carpenter remembers. "He called [a listener] an asshole on the radio. Bob Short came back to me and said 'the guy is over'. I said 'let's give him a break' and he said 'no way'. As it turned out, [Dr. Dave] didn't know the mike was on."

Upon finding out his comment had gone over the air, Carpenter recalls, Dave went on and said something to the effect of "Well I'm suppose to apologize, but I'm not going to, the hell with it". Dr. Dave was replaced with another Carpenter discovery, Wisconsin native B.J. Crocker.

Crocker called himself the "finger-poppin', lipp-smackin', fast-talkin' and even sometimes color-coordenated rock 'n' roll radio announcer". With the popularity at the time of a television show called "B.J. and the Bear" (which was actually about a trucker and his pet monkey), B.J. claimed to have his own "bear", which he called "Goldirocks" and would often tell a story about his "bear", which would end in a punchline.

He was known for his elaborate creativity. He would devise on-air trivia

B.J. Crocker
(Courtesy Larry Wolf)

contests (What group recorded this song? What year was this a hit?) in which the winner would receive a "Doctor of Discs Diploma", which was a poster of the thin, bearded Crocker wearing a doctor's smock and holding the "Buddy Holly Lives" album, or a "Certificate of Smarts" which, printed on the same kind of paper actual certificates are printed on, featured a picture of him standing near the WWTC transmitter tower. He wrote a little message to each of his winners on the posters and certificates and listeners by the dozens would flood the telephone lines to get one.

Brad Piras, the young DJ that Carpenter had interviewed and hired on the roof of the building, was now doing weeknights, after having been a part timer.

Piras had come over from KTWN-FM (107.9), a sedate soft-jazz station based in Anoka. The WWTC format was something completely different but he gradually loosened up.

"I wasn't a real good announcer and I didn't have a whole bunch of schticks", Piras says now. "I tried to rely on music, I tried to rely on getting in and out of things quickly, I just tried to rely on having fun."

After canceling Mutual's "Larry King Show", Carpenter hired Steve "Boogie" Bowman, a San Francisco native known to night owls as "the Nighttime Narcissist", to man the overnight shift. Some strange characters would often come out of the woodwork while Bowman was on and when they called in he would often put them on the air, using them as the element of entertainment on his program. Some of them were heard regularly enough that they earned recognition among his listeners.

The part-time air staff, who could be heard on weekends and filling in for whoever happened to be sick or on vacation, was just as talented as the full-time staff and often became just as popular.

"Ugly" Del Roberts was doing an oldies "cruisin'" show on Saturday nights on WWTC long before a popular FM station in the Twin Cities had ever thought of it, and did it with a lot more authenticity.

He grew up in South Minneapolis and was a DJ at top-40 station KDWB-63 at the age of 20. Now in his mid-30s, Del acted crazier than ever. Speaking at about 90 miles an hour, he talked of cruising Lake Street and other main drags in the metro area. He frequently mentioned the long-since defunct Porky's drive-in, a popular Twin Cities hang-out in the fifties and sixties and would make decidedly politically incorrect jokes and one-line zingers about "good wine and bad girls" and about being ugly. He drew more sympathy than complaints.

"The 'Ugly Del' gag started on the first day on the air and everyone kept calling in to tell me I wasn't ugly", he recalls. "It went on for over a month and the phone calls kept increasing about me not being ugly. I agreed to do a personal appearance with a local rock band on a Monday night and over a thousand of our listeners showed up to see if it was true. WELL. . .!"

Another popular WWTC weekender that Carpenter brought in was Michael Patrick "Records" Ryan, a personality who was familiar to listeners from all over the dial. In the Twin Cities alone, he had worked for KDWB, KRSI, WYOO ("U100"), KTCR and WMIN over the past ten years, as well as in other cities.

Mike Ryan had a dry, sardonic sense of humor and was also known for the jokes and one-liners that one wouldn't get until they thought about it a minute. He constantly identified himself using catch phrases such as "You're plugged in to Records Ryan" and "You're rollin' with Ryan".

Other weekend air personalities included Justin Case, Art Phillips and

Mike "Records" Ryan (Courtesy Stuart Held)

Mary Hatcher. Weekends were special at the Golden Rock. Every weekend had its own "theme", where a particular style of music or fads of a certain era would be spotlighted.

There was a "Surf and Cruise" weekend where Beach Boys and Jan & Dean-type music was spotlighted and a "Wacky Weather" weekend where listeners were to call in when a song that makes reference to weather ("Rainy Day Feeling", "Hot Fun in the Summertime") came on. The first caller to get through won a prize.

One of the most popular feature weekends was the "Copycat" weekend where a song and its hit cover version would be played back-to-back. Other "theme" weekends featured instrumentals, protest music, girl groups and just about any other category the music could be put in.

The disc jockeys made themselves accessible to the listeners. When the telephone request and contest number was given, it was the jock on the air, not an engineer or call screener, who answered.

If somebody called the request line simply because they wanted to talk to their favorite DJ, the DJ would quite often be willing to spend some time with the caller off the air while the music played. The listeners who called in regularly enough to be on a first name basis with the disc jockeys were affectionately known as "groupies". While nobody knew just how many "groupies" there were, there were far more than anyone would likely imagine.

It wasn't uncommon for one jock to walk in on another's show on his own time and joke on the air with him. There was a chemistry between everyone who worked there that was completely unheard of at other radio stations. It was like a big family and the listeners were made to feel part of it.

Fun was a serious business in the eyes of Carpenter. He had a very definite philosophy and idea of how a radio station should sound and how a personality should perform on the air.

"I wanted a perfect performance", he admits. "Every time you opened the mike switch, you were suppose to handle it like this was your last time ever on the air."

"Scott was a music purist or a format purist and he is a control person", remembers Brad Piras. "He wanted to control the way you said something and how you said it. He wanted the perfect half-hour in radio. If there were five seconds of dead air, he would call and be very upset."

On being called a "control person", Carpenter responds, "To control what you said? No. To control your mood and your projection? Yes. But as to the good time you had on the radio, the people you were talking to, the words you used, absolutely not."

The formula was definitely catching on. Fingers were crossed when the Fall 1979 ratings came out. According to the figures, WWTC had a 2.6 percent share of the local listening audience, up a full two percent from the previous book. It was a euphoric moment for a station that had been on the decline for years. But it wasn't enough for the perfectionist Carpenter. He, along with research consultant Bernie Torres of Burns Media, voraciously did their own audience research.

People were defiantly sampling WWTC, they found, and a lot of the samplers were becoming users. There was a lot of positive mail and telephone calls and whenever a DJ offered a prize to a particular caller on the

L-R: **Steve "Boogie" Bowman, Nancy Rosen, B.J. Crocker, Brad Piras and Goldierocks.** (Courtesy City Pages Inc.)

studio line, the lines would be jammed with callers. With the new Arbitron figures, the station began promoting itself to advertisers as "the Twin Cities fastest growing radio station."

Listeners of all ages and genders were tuning in, including teenagers and children as well as people in their thirties and forties, but the listenership was predominantly male. This was perceived as somewhat of a handicap. Advertisers tend to be more interested in women because, according to studies, they tend to buy more goods and are more open to sales pitches. Even Arbitron showed that WWTC was by far more popular with adult men than anyone else.

But the demographic differences were of no surprise to Carpenter and company. As Scott remembers, "We knew we could pick up the male listener first because the male tends to be more of a renegade or revolutionary." He and Torres looked at ways to attract more women listeners without alienating the men.

Parties and Power Struggles

The staff that Scott Carpenter had assembled was young and vibrant. They worked well together because they enjoyed what they were doing and they genuinely liked each other. Unlike other radio operations where cliques and back-stabbing and clashing egos are the norm, everybody seemed to get along. The air staff would party with the sales staff and with the office staff.

A popular hang out was the nearby Court Bar and Grill, where WWTC people could often be seen frolicing. They went to one another's homes and even did some things in the studio and offices. But the partying, consequently, often went too far.

"There was a whole lot of romancing going on", Brad Piras remembers. "More than one of the DJs had been known to have sex while on the air, there was lots of cocktailing going on and other things as well. . .There was a lot of different women that were in the facility and sometimes those women weren't of age. One disc jockey had a tendency to enjoy the young women and cocktails and drugs in large quantities."

One employee is remembered for having pulled a fire extinguisher off the wall and spraying the carpeting in the lobby in several spots.

When the flabbergasted receptionist asked the employee what he was doing, he replied "I'm putting out those fires".

"Fires? I don't see any fires..."

"Then I must have put them all out."

Engineer Al Flom, who had started at WWTC when it was a very main-stream middle-of-the-road music station in the early '70s (and left shortly after Carpenter got in) says "The guys [Carpenter] brought in were dope smokers. Some of the news guys [from the previous format] did smoke dope but they didn't at work. The guys that Carpenter hired were smoking dope in the studios.

"It was the atmosphere that Scott created there that was weird. It's fun to talk about but if you had to deal with these people on a daily basis it got to be a bit much."

Acting general manager Dennis O'Leary decided it was time to put the breaks on the wild activity. He posted a notice in the studio that warned:

"There will be no visitors in the studio after business hours. There will be no drinking allowed in the studio at any time unless it is an official staff function, and there will be no smoking of marijuana or hashish in the studio at any time. ANYONE violating these three rules will be fired immediately."

The night the notice was posted, a WWTC staffer trashed O'Leary's office, overturning plants and knocking over furniture. The staff member was not fired.

O'Leary was only acting general manager, temporarily filling the position after the departure of Red Williams. In November, 1979, the Shorts finally hired Charlie Loufek to fill the position on a permanent basis.

Loufek had been the sales manager at "easy-listening" station WAYL-FM (93.7). Loufek was a congenial older fellow. He was a good communicator and could toe the line between the owners and the staff. He was generally liked by everyone.

Dick Driscoll (Courtesy Larry Wolf)

Loufek had his own ideas on how to run the station. In December, he hired Dick Driscoll (without consulting Carpenter, much to the program director's dismay), who was an announcer at WAYL. While continuing at WAYL, Driscoll began hosting a late-afternoon air shift on WWTC and was appointed operations manager.

He was one of the area's early rock 'n' roll disc jockeys. His thick, baritone voice had been heard on WDGY in the late fifties and early sixties and was familiar to most local radio listeners, having been heard around the dial, encompassing almost every conceivable format as well as being the voice-over in numerous commercials. He had been involved in almost every aspect of broadcasting, including engineering, management, programming and sales, as well as announcing.

Driscoll was no stranger to WWTC. He had been a program director there for a time, and a personality from 1969-1975, with a different format, different management and different owners. He had also worked in the same downtown Minneapolis studios ten years before that, when they were occupied by WDGY. (WDGY moved out and sold the facilities to WWTC in 1964.)

Driscoll was slipped into a late-afternoon time slot, between B.J. Crocker and Brad Piras. Having played the sedate "elevator" music of WAYL for the past eight years, he welcomed the opportunity to let loose and rock 'n' roll. He quickly became popular with listeners, many of whom remembered him from WDGY 20 years earlier (as himself and as his alter-egos: "Count Dracula", "Maintenance Man" and traffic reporter "Flying Officer Nelson", all of whom found new life on WWTC).

Carpenter and Loufek got along fine on a personal level but Carpenter began to get the feeling that Loufek was trying to push him and the consultants out. Both men were strong-willed and somewhat egotistical and a battle for King of the Hill seemed to develop.

Carpenter was particularly annoyed at the hiring of Driscoll. Not that he had anything personal against Driscoll, but in his view, it was him, not Loufek, that had the responsibility of hiring and firing staff. Carpenter and Loufek became a bit wary of each other and began to do things behind each other's back to protect their own perceived positions of authority.

Meanwhile, Driscoll, as operations manager, was himself trying to bring new staff in the door. His young protege Arne Fogel, dubbed "the World's Greatest Trivia Expert", began turning up on the airwaves as Driscoll's frequent in-studio "guest", rapping about music with the host and callers.

With Fogel's gigantic collection of records and his vast knowledge of music, Driscoll wanted to bring him in as a program producer and possibly music director. But Carpenter wouldn't hear of it and Fogel was kept an outsider — for the time being.

Purple Paint On the Walls

"If Muzak is aural wallpaper. . .", observed Deborah Miller of a weekly newspaper called *Sweet Potato* (now *City Pages*), "WWTC is purple paint on the walls. It intrudes, it makes you pay attention, it yells at you to start dancing, it makes you reach for the phone and it never lets up."

One couldn't listen to the Golden Rock passively; it made a point of jumping in your face and keeping your attention. One minute you might hear a song that might strike up a memory of hanging out with the guys or making out in a backseat or cruising around in your first hot rod; next minute you'd hear Ugly Del calling a telephone located in a women's public restroom on the air or Dick Driscoll, portraying his infamous "Maintenance Man" character, deliberately causing a "technical difficulty" or "Records" Ryan following up a joke with the sound effect of a toilet flushing.

"The early days [of the Golden Rock] were the best, there's no doubt about it", says Mike Ryan. "When people were around the station they were all having a good time. It was like the object was to just have fun and go out and do it. It wasn't uncommon for somebody to come in during their off-hours just to say 'Hi, how are things going.'"

Adds Nancy Rosen, "[It was] great fun. It was so great playing the oldies and having this free type format. It makes the job exciting. You don't have to do liners, well we had some, but other than that it was just free and that made it fun. We always did requests and we'd put them live on the air. . .I talked to millions of people and it just made it fun because you're in touch with your listeners and that's unheard of in radio now."

If WWTC had anything going for it, it was a tremendous amount of listener loyalty that was virtually unheard of at other radio stations, even then. The audience was encouraged to participate in their radio station and they faithfully did, checking in with the DJs regularly, participating in events sponsored by the station and even contributing to the music library by bringing in their own records and allowing them to be taped for the station's permanent collection. If a disc jockey on the air said he was looking for a particular song, someone would inevitably bring it in.

A local collector named Bob Broz, with his thousands of records, made so many contributions to the station that he would often be an in-studio guest, usually with B.J. Crocker, playing obscure recordings and talking about the songs.

The station had only a shoestring budget for promotions. All the advertising and contests were created in-house because the owners didn't want to spend money on an outside ad agency. But the staff was shrewd enough to come up with some clever and occasionally outrageous gimmicks to capture the attention of the public and sometimes the press.

Contests on the Golden Rock could often be as odd as anything else on the station. Usually it was just a simple trivia contest; what year did this song come out, who recorded this one-hit-wonder, etc., but sometimes 'TC listeners were coaxed into doing some off-the-wall stunt just to win a cheap prize.

B.J. Crocker had "caller number ten" sing along on the air with the Monotones' 1958 hit "Who Wrote the Book of Love" for a prize one afternoon and had callers guess how many times the word "bird" is repeated in "Surfin' Bird" by the Trashmen on another afternoon.

Some of the contests were a little more outrageous. One day, WWTC went to the beach — in the dead of winter.

The staff came to work wearing shorts, bathing suits and Hawaiian shirts, carrying surf boards and beach towels. Listeners were invited to come down and the first one to show up in a bathing suit won a "poor-man's trip to Hawaii" — a grass skirt and plastic pineapple. Dozens showed up in beach attire while snow covered the ground and wind blew outside.

Recalls Scott Carpenter, "We went down the street and did stuff and people were freaking out. We got a whole bunch of listeners to put on bikinis and there were ladies in bikinis in the middle of cold winter coming down and we all went up to that roof site where we took a big photo of the WWTC Day at the Beach in the middle of winter. And here it is, the steam coming out of the chimneys and everybody's breath is visible and here's the women in these gorgeous two-piece bikinis standing out there in the cold. I mean, it was so great."

For their troubles, Carpenter bought everyone dinner. "Thank God Bob Short owned the Leamington Hotel, with the Black Steer restaurant."

In the summer of 1980, station employees formed their own softball team, the "TC Dugout Disasters", which played the employee teams of other radio and TV stations. (One of the few games the Dugout Disasters won was in August, against WCCO-TV).

One of the biggest practical jokers at WWTC was "Ugly" Del Roberts. Often creating his own stunts, among the memorable examples was his (unauthorized) chocolate chip cookie contest.

"It was nothing more than a scam to mooch free cookies which I loved", he remembers. "I offered some silly prize. Nobody approved the contest or knew about it. The next thing I knew, cookies started to arrive. Bags and bags of cookies. Famous restaurants sent in entries. The bags numbered into the hundreds. Of course we then needed milk so we mooched milk from the listeners."

The on-air stunts kept people listening closely and just like on the old "Candid Camera" TV show, one never knew if they might inadvertently become part of a stunt. Resident prankster Roberts often used the telephone to "entrap" people, such as the time he called a pay phone located in a women's rest room at an Ember's restaurant, asking the incredulous passerby to dedicate the next song.

"This is the bathroom", a woman responded, having been informed she was on the air.

"I know that. Would you like to dedicate the next song?"

"You're crazy!"

On an afternoon when he was filling in for an ill B.J. Crocker, Roberts said B.J. had "kissed his bear and bears carry diseases". The afternoon was filled with

WWTC personalities hosted events at nightclubs such as Sam's (now First Avenue).

The Incredibly Ugly Del Roberts (Courtesy Larry Wolf)

reports on the supposed whereabouts of Crocker, which included being chased by zookeepers at Como Zoo and auditioning for a Hamm's beer commercial.

He also called a local McDonald's restaurant, live on the air, to satisfy a curiosity that had been bothering him.

"McDonald's, may I help you?" asked the perky young girl on the line.

"Hi, this is Ugly Del Roberts calling live on the air on WWTC."

"You're kidding!"

"No, not at all. I was wondering how many billions of hamburgers have been sold. I haven't seen the sign lately."

"Ah. . .let me get the manager."

The restaurant manager picked up the phone and after hearing Del's inquiry, gave Del the phone number to McDonald's corporate offices, rather than just looking out the window at the sign.

Still on the air, Roberts called McDonald's corporate offices and after being switched to a few executives who had no idea how many burgers had been sold, he was finally given the latest exact figure by a bemused McDonald's executive.

Everyone had their own on-air irreverence. Weekend DJ Mike Ryan would say he played "favorites — MY favorites" adding, "if I don't like it, you won't hear it."

His zingers occasionally bordered on the risquc, such as when he played a 1966 hit by the Swinging Medallions and said "I too need a 'Double Shot of My Baby's Love' — both barrels if you get my drift!" His show was often punctuated with the sound effects of laughter, flushing toilets and recorded comments, such as the teenage girl who sneered "That's Tacky!" and the young man commenting, "That's the dumbest thing I ever heard of!"

For Dick Driscoll, it was like reliving his youthful days some 20 years earlier on WDGY. Not only was he playing the same music and doing some of the same stunts he did then, but he was doing it from the same rooms, as the studio itself was once occupied by WDGY.

"I had been at [easy-listening station] WAYL eight years", he said in a *Minneapolis Tribune* profile. "But after a couple of days at WWTC, I started saying 'What am I doing? Am I trying to relive my youth or what?'"

Nancy Rosen, the youngest of the air personalities was billed as the "sweetheart" of the staff. While she was often portrayed as the little sister tag-along, she was also the lone sane person in a ship of fools.

"I was the more mellow of everybody and I don't want to say this in a bad way, but more real. I talked with my listeners more than I joked with my listeners. I've always been an information giver of the music. I love what I call 'worthless trivia' so my show always had worthless trivia on it."

Sometimes an actual predicament that a disc jockey found himself in took on a life of its own at WWTC.

Early one morning, Scott Carpenter was driving to work in a car that had been leant to him by a local Pontiac dealership that advertised on 'TC. The sales director cut a deal where, in exchange for commercials, the dealership leant out new models to some station staff and management.

The vehicle Scott was issued was a particularly nice one, a black, fully-loaded 1980 Grand Prix. While driving it into downtown Minneapolis, he waved to some police officers he thought he knew. The officers didn't think they knew him and became suspicious at his seemingly cavalier attitude.

The next thing he knew, he was surrounded by squad cars, lights flashing. He was ordered to get out of the car, hands in the air. One of the officers he had waved to had checked out the license plates and found that the car had been reported stolen. Carpenter was handcuffed and escorted to the back of one of the squads.

Shook up and confused, he tried to explain to the officers that the car was leant to him by the dealership. After a while the cops began to believe him and finally contacted WWTC's sales director, at home and in bed, who corroborated the story.

As it turned out, the vehicle Scott was driving was leant out by mistake. When it was discovered missing from the lot, it had been reported stolen. The handcuffs were finally removed, apologies were made and Scott was on his way.

Meanwhile at the radio station, Nancy Rosen was doing the morning show herself with no idea what had happened to her partner. Carpenter showed up 45 minutes late and explained his tardiness to her on the air.

He then had an idea. Since he had friends on the force, he managed to contact the same officers who had almost arrested him that morning and persuaded them to go on the air with him. Having a sense of humor about the whole thing, they kidded with him about the incident. Suddenly cops from Minneapolis and St. Paul were calling in, teasing him on the air about his predicament.

The WWTC hot air ballon hovered over the Twin Cities in the summer of 1980.

Before the morning was over, Carpenter managed to get the Minneapolis Chief of Police to stop by the studio. The Chief announced on the air that Scott was, in fact, under arrest after all. The charge: impersonating a radio announcer.

Scotty Beams Up

Scott Carpenter had fun at WWTC but as the year turned into 1980, he was becoming more and more frustrated. He brought the station up from the ashes but he could only bring it so far. The obsessed perfectionist was suddenly stagnate. Scott realized all too soon that the budget, the ownership and the politics limited the progress that the radio station could ever make, despite the potential.

"Here I'm trying to make a station get into the four to five share range and I've got no budget", recalls Carpenter. "Bob Short really didn't have the cashflow. It was a big deal when we got $5,000 to spend. All I could do is try to get a good format together, a good station but I had no real incentive to stay there because it never would be promoted correctly."

Indeed the extent of the station's promotions, outside of on-air contests, were a few billboards, an infrequently seen, low-budget TV spot and bumper stickers given away at just two locations, one being the WWTC office. Other stations could afford to spend millions branding their call letters into people's minds, saturating them with commercials, billboards, newspaper ads and direct-mail campaigns. They could also afford to hire an ad agency and not have to create everything themselves.

"You just can't sit there and do the job and hope people will come and listen to your station. It doesn't work that way", Carpenter says.

Carpenter had been negotiating with WCCO Radio for sometime. He grew up in Minnesota and enrolled in Richard Brown's nationally famous broadcast college after high school, aspiring to one day work for the "Good Neighbor to the Northwest".

By 1964, at the ripe old age of 20, he was announcing at WLOL (1330) in St. Paul. The following year he was spinning records for St. Louis Park-based KRSI (950) and by the end of the sixties he was a "boss jock" at Duluth's top-40 powerhouse, WEBC. After bouncing around from city to city throughout the seventies, he hoped his return to Minneapolis would be a stepping stone to the station of his dreams. It didn't quite work out that way. He was a little too creative for this market.

Carpenter was a man of contrasts. People either loved him or hated him. He could be friendly and encouraging, and he could be demanding and difficult.

"He was really tense, he has a really intense personality", says Nancy Rosen. "He's brilliant, but a little too off-the-wall. But he knew exactly what he was doing, he did a great thing with the radio station for the short time he was there."

Observes Dick Driscoll, "Most radio personalities have some strange quirks and certainly Scott had a few of his own".

He had unusual ways of making a point, such as when he called a staff meeting to order crushing a styrofoam cup full of ice and soda-pop in his hand. After ice chunks flew all over the room he said "Now that I got your attention. . ."

He was often described by both friend and foe as "weird".

"I don't think weird is wrong", he says. "To be program director, to be creative, you have to be outspoken. You have to do things that are extroverted."

His creativity and outspokenness perhaps worked against him in a city where a lot of people are scared to death to make waves or possibly offend.

Scott Carpenter left WWTC and Minnesota for good on April 20, 1980. He returned to Florida, where he had been working before coming up north and eventually migrated to the west coast, switching to talk radio and hosting a call-in program on the CBS-owned station in San Francisco and later for ABC Radio in Los Angeles. He was also the official voice for the California State Lottery through the 1980s. He never set foot in Minnesota again. With his departure, WWTC severed its relationship with the consulting firm of Burns Media as well.

His young protege, Brad Piras, inherited the title of program director and took over Carpenter's morning air shift. He realigned the schedule, slipping Nancy Rosen into the evening time slot (research showed she was a lot more popular with the young male listener, who was more inclined to tune in at night than with the older females who listened during the day). The Mighty Dick Driscoll was on from 9 a.m. to 2 p.m. and B.J. Crocker began working afternoons from 2 to 7 p.m.

Ironically, shortly after Carpenter's departure, the Spring 1980 Arbitron ratings were issued. WWTC had acquired a full 3.9 percent of the listening audience, up from 2.6 percent six months earlier. It was the third-highest rated AM station in the Twin Cities, and the second most popular station, AM or FM, among 25-49 year-old males. The staff celebrated with a massive, wild party. But they didn't have the insight Scott Carpenter had.

Stump the Chump

The week Scott Carpenter left, Arne Fogel, the frequent in-studio guest of Dick Driscoll, was hired to produce and narrate weekly tributes to popular music artists, for broadcast on Sunday nights.

His first show spotlighted Bob Dylan. Fogel played both familiar and not-so-familiar Dylan music with anecdotes and passionate commentary on his life and career. The response was tremendous.

In the ensuing weeks, he did tributes to the Beach Boys, George Harrison, the Who and even a tribute to old TV themes. The programs seemed to have the professional quality of a syndicated show.

Soon Fogel was doing odd air shifts at the station, usually midnight to 6 a.m. on a weekend. But he would also be playing a bigger roll there.

While most weren't familiar with the name, Arne Fogel was something of a fixture in Twin Cities media. Since 1968, at the age of 19, he worked as a session musician and as an advertising writer and jingle singer. Throughout the seventies he wrote and sang in numerous familiar commercials, including ones for Ember's restaurants, Slumberland furniture stores, Arctic Cat snowmobiles, Gaymont yogurt, Hormel hot dogs and Minnesota North Stars hockey.

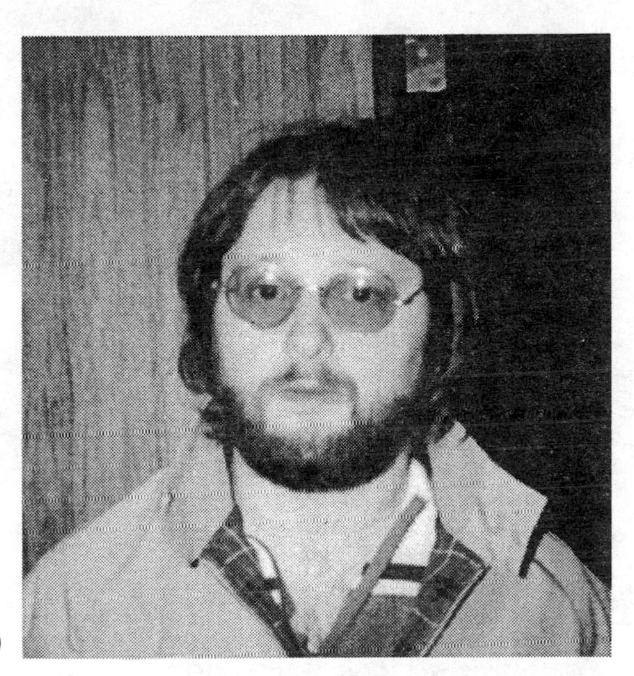

Arne Fogel
(Courtesy Larry Wolf)

He had also been in some local rock bands in the late sixties and early seventies, notably thc Batch, and the Puddle.

When WWTC general manager Charlie Loufek found out about Fogel's advertising background, he offered him a job in the sales department, working with the sales staff, writing copy, and producing commercials for clients that had no agency, in addition to his on-air duties.

"I just grabbed at that", Fogel says. "It was still part time, but it was like every day for half-days. So if you listened to the station around 1980, you got a sense of me being there a lot more [than I was]".

Fogel would often walk in while others were on the air, chatting and joking with the DJ almost every day, gaining notoriety with the listeners even though he only did obscure weekend shifts and Sunday night "tributes". "I was in the back room doing commercials all day".

On air, the soft-voiced, intellectual Fogel wasn't the stereotypical DJ.

"Brad Piras tried like hell to make a rock 'n' roll disc jockey out of me and I just didn't have it. Male rock 'n' roll DJs have a certain kind of 'macho strut' and I just didn't have those kind of pipes".

Thursdays at noon, Fogel joined Dick Driscoll in something called "Stump the Chump", where Fogel, who was billed as "The World's Greatest Trivia Expert", would challenge listeners to call in with an entertainment-related trivia question. If the caller stumped him, that caller won a WWTC T-shirt or tickets to an event sponsored by the station.

"Dick and I patterned our relationship on the air after Jack Benny and Dennis Day", he says. Arne, playing the roll of Day, would be "the kind of dumb guy who occasionally was the wise-ass", the "foil" to Drscoll's "straight man". Driscoll would often use Benny's old line "Now you cut that out!"

Dedication Lines Are Open

Listener participation was an ever-important element of the Golden Rock. Listeners were always encouraged to call the jock on the studio line to make a request, dedication, to play a contest or just to talk. WWTC acquired quite a legion of loyal listeners as a result.

Something odd was always happening at 'TC. "Nighttime Narcissist" Steve "Boogie" Bowman once offered a watch to the listener who could bring Brad Piras, whose show succeeded Bowman's, in to work. Piras was promptly "kidnaped" by a faithful listener and brought in. Piras himself was known to have coffee-slurping contests on his morning show where listeners would call in and literally slurp coffee over the phone, on the air, as well as hog-calling and yodeling contests.

When a female caller said something complimentary to Mike Ryan, he'd respond in mock bashfulness, "Are you talking to me?" Dick Driscoll, in the persona of "Count Dracula", was known to sing "A tisket a tasket, a green and yellow casket".

Of course the Golden Rock wasn't mainstream and those who were not fans had a hard time understanding those who were.

"My girlfriend at the time and I used to argue over that damn radio", recalls an avid 'TC fan who was in high school when the Golden Rock was popular. "I always wanted 'TC on but she insisted on listening to [top-40 FM station] WLOL. She thought I was crazy and I thought she was."

At the 1980 Aquatennial parade in downtown Minneapolis, Arne Fogel, weekend DJ Art Phillips and traffic coordinator Sandy Charles rode in the "TC Golden Mustang" waving to the crowds when a young woman ran up and shrieked "WWTC! I love your station!"

"Great!" called back Arne.

"My friends think I'm NUTS!"

And maybe a few 'TC listeners really were "nuts". Dick Driscoll recalls what he describes as a nerdy young man who came to the studio, tore a WWTC bumper sticker in half, held up the "TC" part and said, quite seriously, "I like this part, I don't like the other part."

"Okay, why don't you like the 'WW'", Driscoll asked.

"World War. I don't like that."

The disc jockeys often poked fun at the songs they played, none more than "Ugly" Del Roberts. He'd dedicate the Gene Pitney hit "I'm Gonna Be Strong" to "those of you who forgot your deodorant this morning" and Johnny Preston's "Running Bear" to "all you nude joggers".

The theme from "Shaft" was dedicated to car dealers and "Ring of Fire" went out to those with hemorrhoids. "I did get some flak when I played 'Love Hurts' for all the virgins", Del remembers.

Listeners were invited to make their own dedications. When the dedication lines were open, people would call in and dedicate the upcoming song to someone, live on the air.

Typically, the DJ would answer the phone on the air, saying "Hi, 'TC, who do you want to dedicate this to?" and the caller might say "I'd like to dedicate this to Susie from Bob" or "I wanna dedicate this to my friend David and my step-cousin Tom". About four or five calls would be taken and the song, usually something nice but not mushy, would be played in honor of everyone mentioned in the succession of calls.

Of course, going directly on the air with no tape delay or any other way of bleeping anything out had its share of risks. Although it didn't happen often, there was occasionally someone who tried to say a profanity on the air or pull some other stunt.

On a Saturday night when Ugly Del was doing dedications, a precocious youngster came on the line.

"Before the babysitter catches you, who do you wanna dedicate this to?" Del asked.

"KDWB", replied the high-pitched voice.

"Thank you." Del then hit a button and a recorded voice retorted "Aw, stick it in your glove compartment!"

Being the main female voice on the station, Nancy Rosen was occasionally prone to the rude caller and she is kidded to this day for one on-air incident.

"Let's take some dedications for this next song", she said in her cheerful voice.

"Yeah, I wanna dedicate this to my uncle who lives in Georgia and my cousin Betty-Ann."

"Alright, you got it! Next, you got a dedication for us tonight?"

"Yeah, I wanna dedicate this to Sue."

"Okay you got it!"

A pleasent-sounding young man followed on the air. "I wanna dedicate this to you, Nancy, 'cause you're the best disc jockey on the air!"

"Well alright, thank you!"

"And you wanna know something else, Nancy?"

"What?"

"I really would like you to sit on my face."

Without missing a beat, Rosen introduced the song, acting as if the remark didn't phase her. But when Arne Fogel came in to work 20 minutes later, she was visibly disturbed. When he greeted her, she said, "Did you hear that?!" But her distress didn't come out on the air.

"I was still fairly a rookie in the business, only a few years under my belt, and that was my first time ever really doing that phone stuff", she says. "We didn't have seven-second delay and you just don't think you're going to get calls like that."

"You can't ignore nervously", Fogel says about the smooth way Nancy handled the situation. "You have to ignore like it was in the script".

People who called the disc jockeys, both on and off the air on a regular basis were affectionately referred to by the air staff as "groupies". While many of the "groupies" called all the jocks, each DJ had his or her own mini fan club.

"We had groupies by the ton!" recalls Del. "One chased Brad Piras for weeks. Even Nancy had problems with male listeners. I think we all yielded to the pressure. It seemed like WWTC was all party, even on the air."

Steve "Boogie" Bowman, who did the graveyard shift, had his own slate of regulars who came out of the woodwork after midnight. There was the young woman who called herself "Laser- Brain", the high school kid known as "Gillman the Land-Mass Collector", there was "Yacky Duck", "Little Itty Bitty", "Jumpin' Joe the Mass Dedicator" and "Mr. Lonely"; characters Larry King no doubt would have hung up on if 'TC ran his show (which still came in on the network feed) instead of Bowman's.

A lot of interesting people would call in during the wee-hours to talk to the one friend they had that was up so late at night. Arne Fogel, who often did the overnight shift on Friday and Saturday, recalls a seductive young woman who would administer phone sex to the unsuspecting disc jockey.

"She used to. . .she was. . .whew! She used to talk about her body", recalls Fogel. "It's three in the morning and you're not terribly interested in what you're doing, you tend to pick up the phone during songs and say 'So what's new now?' and she'd say 'Well right now I'm playing with my. . .' and so you kinda go 'Oh yeah? Okay."

The mysterious woman would frequently call Bowman as well. She would say with a lot of breath in her voice "I'm sooooo slender. I'm so sexy. I'm built just the way a man likes."

Ugly Del hosts the second 'TC class reunion while Nancy Rosen dances to the music. (Courtesy Del Roberts)

Nancy Rosen talks to some fans at the 'TC class reunion. (Courtesy Del Roberts)

"You knew after a while she was probably just the opposite of that and needed to have somebody to believe that that's the way she was", speculates Fogel. Bowman once invited her to the studio, out of genuine curiosity about the mysterious woman, but she never did show up.

The World's Greatest Class Reunion

The relationship the audience had with the Golden Rock went beyond tuning in the station or calling the disc jockey on the request line. The 'TC staff was well known for partying, so it seemed only fitting to throw lavish parties for the loyal listenership.

The parties were called "class reunions". Those who attended were encouraged to dress in their high school fashions. There were plenty of poodle skirts, zoot suits, high tops, hip huggers, bell bottoms and saddle shoes worn to the events.

Described in a nutshell by a 'TC fan as a place of "high energy, high volume rock 'n' roll", the Class Reunions were listener appreciation parties where Golden Rock fanatics could meet their favorite WWTC personalities and connect with one another. That aspect was important when the average Golden Rock fan was usually looked upon with disdain by their FM-listening contemporaries. Everybody had a good time as they danced, drank, shook hands, exchanged phone numbers and partied.

"It was just people having a great time and these were our people", says Nancy Rosen.

The events were something of a "happening"; for a five-dollar ticket, attendants got to see performances by several rockabilly and "oldies" bands such as the Rockin' Hollywoods, Hitz, Flash Cadillac and the White Sidewalls, consume plenty of refreshments, dance in the giant ballroom and mingle with their favorite WWTC personalities, who were dressed to the hilt. The events were broadcast live on the air at 1280.

The first Class Reunion was held June 1, 1980 at the Radisson South ballroom on I-494 and Highway 100. The station originally intended on booking it at the Leamington Hotel in downtown Minneapolis, which along

WWTC gave its air personalitites more prominence than its music, although Nancy Rosen's name is misspelled (not to mention that she was a long way from graduation in '65!)

with WWTC was owned by Robert Short. But the hotel, figuring it wouldn't make much profit on the event, turned it down. The Radisson South booked it and all 2,000 tickets were sold or given away well in advance. Although the event was sponsored by the local branch of Olympia Brewing Company, Del Roberts, who emceed the event, recalls that "every drop of beer in the building was gone in 45 minutes".

Observes Rosen, "This is what was so cool, you knew 2,000 people were there because they were a fan of the radio station. The thing that really thrilled me was to see these parties sell out like that and know that every one of those people in that room loved WWTC."

Upon realizing the blunder of refusing the first Class Reunion, the Leamington Hotel took the next one, held September 21, 1980, with open arms. The holiday class reunion also sold out although it was held the day after Christmas.

Along with the extravagant listener appreciation parties, WWTC came to private parties as well with its "Sound of Music Machine" entertainment service. The service provided not only music, but an actual 'TC disc jockey as host.

The "Music Machine" was a mobile unit designed and equipped to provide music and a WWTC personality for dances, class reunions, parties and other gatherings. It came complete with amplifiers, loud speakers, microphones and even special lighting, all provided along with the 'TC personality for one package price. When the "Sound of Music Machine" was hired, a colorfully painted WWTC van would show up, carrying all the equipment. The music provided was on the same cartridges used on the air (literally) and the "Music Machine" client could ask for music from a specific year, era or a mix of '50s, '60s, and '70s music. As an added feature, "The Golden Rockettes", an all-female dance group could be hired for special events, performing in various cheerleader outfits, poodle skirts, etc.

According to Dick Driscoll, WWTC's "Music Machine" was more successful than the party service provided by WLOL-FM (99.5), the contemporary hit giant of the time, even though WLOL had a light show and better equipment. The reason, quite simply, was that WWTC sent out one of their own air personalities to handle the gig. WLOL would contract somebody else to do it. It meant a lot to have that "celebrity" there at a birthday party or sock hop or class reunion. When WLOL came, the person sent out wasn't someone anyone "knew".

WWTC also paid its disc jockeys fairly well for the "Music Machine" gigs, more than the much bigger WLOL paid. A 'TC jock made more money in one night doing a Music Machine than they made in almost a week of regular airshifts on WWTC, as the money was coming out of someone else's pocket besides Bob Short's.

Local Gold

The Golden Rock was geared toward the listener who grew up in the Twin Cities. The disc jockeys often talked about the popular hangouts of yore and mentioned high school classes of the past. The music played was based more on what was locally popular than nationally popular. Those responsible for putting the music library together researched the old KDWB "6+30" weekly music surveys that the station used to give out more than past issues of *Billboard*.

In the field of popular music, certain songs may be big hits in one city but not in another. In an eclectic market like Minneapolis-St. Paul, which has been and continues to be a hotbed for locally based bands and recording studios, a lot of records are released locally, given some airplay and sold regionally but never make it into *Billboard's* "Hot 100". KDWB and WDGY, once the AM top-40 giants played many of those records in their time but they had all been pretty well forgotten — until WWTC began giving them a spin.

Along with all the big hits from Elvis and the Beatles and everyone else, 'TC played "Because of You" and "The Grind" by Gregory Dee and the Avantes, "What Is the Reason" and "Let the Good Times Roll" by the Delcounts, "Midnight Train" and "Feels So Fine" by the Castaways, "King of the Surf" by the Trashmen, "Action Woman" by the Litter, "Cottage Cheese" by Crow and many others.

The concept of playing local oldies originated with Ugly Del, who happened to have the records in his own collection. Del had a hard time at first convincing management to add those tunes to the permanent library. He contended that the music was important and that it would be foolish to leave it out. Management didn't want to take the risk.

Del began playing a few of the local tunes on his own show. Much to the surprise of management, the response was tremendously positive. Listeners were calling in and requesting the songs, and asked about other local hits.

Management agreed to add a few of the songs to the permanent collection and when they were played, listeners called in and asked for more. Del did entire programs of local music, talking between the records about the clubs the bands played in around the cities, such as Mr. Lucky's and Magoo's.

One afternoon, Del played an obscure novelty record from the early seventies which told the story of a young man who must arm wrestle a muscle-bound, beer chugging oaf to win the heart of a girl. Included in the song were references to Mayslack's bar, Elsie's restaurant and other Northeast Minneapolis hot spots. In the ballad, the hero loses the arm wrestling match and has to "play the organ next Sunday when my best girl marries that jerk". He assures, however that he'll only play half of the song "cause my left hand is all that will work".

The song was called "Sven Ivan O'Myran Wisnewski", the name of the singer's nemesis. Del had a contest offering a prize to the first person who could name the singer on this record.

The singer of this rather silly song from 1972 was none other than a young Arne Fogel. The record had been released originally as a promotion for Northeast State Bank of Minneapolis. Long before becoming part of the WWTC family, Fogel had been (and continues to be) a respected studio musician and singer.

Listener response to the playing of "Sven Ivan" was tremendous enough that it not only drew requests, but people were calling in and writing for copies of the recording. Soon, the single was reissued on the locally-based Maplewood label and Fogel found himself autographing record sleeves at a downtown Minneapolis record store.

Another obscure record Del discovered and played on the air was of a guy playing piano while belching in perfect harmony. "It was the grossest thing ever recorded but the guy's family heard it and called in. It had been done years ago and they forgot about it and didn't think a copy of the record still existed."

Mutual Climax

WWTC was the Twin Cities Mutual Broadcasting System affiliate during the Golden Rock era. Distinguished as the only major radio network to not venture into television and the one network usually identified by name rather than initial, the Mutual network was formed in the mid 1930s as a co-operative between three stations: WOR New York, WGN Chicago and WLW Cincinnati. As other stations joined, they became "partners" in the ownership.

By the time WWTC joined the network, it was a wholly-owned subsidiary of Amway and the largest single radio network in the country, with over 600 affiliates nationwide.

As the local Mutual affiliate, WWTC was obligated to carry five minutes of network news at the top of every hour and a five-minute network sports update daily at 1:35. For the DJ, that meant the song he put on just prior to the top of the hour had to be timed just right. If the song went over-time, the DJ was torn between cutting into it for news, which would irritate the listeners who would bombard him with calls for the next ten minutes or joining the network in progress, which wouldn't make the folks at Mutual very happy.

WWTC did not carry all of the programming offered by Mutual, including the popular "Larry King Show" (which was given to KSTP Radio) or network coverage of NBA basketball and Notre Dame football. This often resulted in some awkward situations for the DJ on the air.

Early one morning, as Ugly Del was filling in on the overnight shift, he fell asleep during Mutual news. Leaving the network feed on, Larry King came on and it was another fifteen minutes before Del finally woke up and got the music going again.

Play-by-play sports coverage often preempted the hourly news and almost every 'TC jock at one time or another cut into the middle of a game, expecting news. This could create a very embarrassing situation for the jock, especially if he had to do something (such as go to the bathroom) during the five-minute break. Avid sports fans Brad Piras and Mike Ryan would sometimes listen to the games in their headphones and give score updates while they did their shows.

Another problem was, when the news was on, it was often fed to the station a few seconds early or late, which could be frustrating for the DJ, who tried to time everything to the exact second. "I think Mutual went by a different clock than the rest of the world", says Dick Driscoll.

"After I got burned a couple of times, I learned real early to monitor", says Mike Ryan. "[Mutual] always had a beep ten seconds before they did the news. . .I learned to listen for that ten-second warning beep instead of looking at the clock."

WWTC International

WWTC had a respectably large listening audience in the Twin Cities and it was easy to find on the AM dial, but it was no powerhouse.

The station was licensed by the Federal Communications Commission to operate on a 5,000-watt directional signal. They could operate at full power during the day, but at sunset the station had to redirect its signal to keep it from colliding with other stations on the same frequency. As a result, the WWTC signal would often shoot clear across Lake Superior and keep on going on clear nights. The nighttime disc jockeys often got requests from listeners calling all the way from Canada but the signal went even further than that.

Many people all over the world, make a hobby of "DX-ing", picking up distant radio signals and sending cards to the stations they pick up with information on what they heard and at what time. The cards are stamped for verification by the radio station and returned.

According to operations manager Driscoll, "We got letters from Sweden, Norway, places like that, saying what time they heard Nancy or B.J. on the air with the Golden Rock, the name of the tune they were playing, things like that."

Driscoll estimates WWTC's listenership in Scandinavia to have been at about 100 but it was inconsistent. The signal didn't travel that far every night and it bounced around in other directions as well, depending on the atmosphere.

Eakman Vs. WWTC

Sunday mornings from six to twelve noon when WWTC wasn't playing the Golden Rock, the station aired religious and public affairs programming. One of the programs, a live broadcast of services from Edina Baptist Church aired on the station under the title "Encounter".

"Encounter" was often more than a Sunday sermon. There were political discussions and endorsements, and listeners were encouraged to lobby and vote on issues such as abortion and the Equal Rights Amendment.

The mixture of political and religious issues prompted a Minneapolis resident named Marvin A. Eakman to sue the church and WWTC Radio on the grounds that the program violated the constitutional separation of church and state. Eakman charged that the church had no right to discuss political issues in a religious program and that the radio station was liable for airing the program.

The suit was heard in U.S. District Court by Judge Edward J. Devitt on May 5, 1980. Brian Short, son of station owner Robert Short and an attorney by profession, served the defense and called the complaint "frivolous", arguing that such a complaint should be heard by the Federal Communications Commission and not in United States District Court.

Judge Devitt agreed and dismissed the suit, telling Eakman "If it really bothers you that much, why don't you just switch to another station?"

Eakman contended that there was more at issue than just changing stations. He told the *St. Paul Dispatch* that the whole issue involved a principle that could "affect millions".

Eakman vowed to take his complaint to the FCC but nothing came of it. "Encounter" continued on WWTC, with as much political discussion as ever. Some 'TC staffers secretly hoped Eakman would prevail, not so much because of the politics, but because the live broadcast of the church service aired on the insistence of Short, against the better judgment of staff who felt the station would do better playing the regular rotation of music on Sunday mornings.

The Iron Hand of the Shorts

WWTC seemed like a small, close-knit family but it was just one of many businesses owned by Minneapolis businessman Robert Short and his seven children. The Shorts had a reputation for running their businesses tight-fisted and as far as they were concerned, the radio station was just another part of the empire, along with the trucking firm, the hotels, the parking lots, the office buildings and anything else they owned.

Bob Short bought WWTC from New York-based Buckley Broadcasting in 1978 for $600,000. Then a news-talk station, WWTC was Buckley's weakest property. Short, who was running for Senate at the time, left his attorney son Brian in charge.

The relationship the Shorts had with the radio station they owned tended to be a stormy one. According to Driscoll, "[Robert] Short hated that radio station. He bought it during his campaign as he was running for Senate. But the station ran editorials against him. . .he hated that station and just didn't want anything to do with it."

After the staff completely turned over and the format changed, Short was still wary of the black sheep of his empire. "It was six months after I got there before we could convince him to come over and see the radio station", recalls Driscoll. "He finally got to like the thing, he became proud of it as his station."

While WWTC surprised everyone by becoming the fastest-growing AM station by doing everything that's not suppose to work according to so-called "experts", the Shorts apparently didn't understand the necessity to nurture what they had.

Observes Driscoll, "By 1981, that station was making money hand-over-fist. But instead of paying it to the staff, instead of dumping it back into the station for equipment or promotion or whatever, that money was [used for] paying off [the purchase price of the station] and everything else."

Former 'TC staffers make no bones about admitting they were paid an average of $3.50-$4.50 an hour. Most of the equipment in the main studio was the same equipment that was there when Driscoll worked for WWTC some ten years earlier. And promotions, such as advertising and contests, came as cheaply as possible.

The Shorts often did consider improving the station's facilities, however. They reportedly looked into buying station KTCR-FM (97.1) for a mere two million dollars, and possibly simulcasting 'TC's Golden Rock programming on the FM signal. The Shorts passed it up, however, and KTCR was eventually sold to Parker Communications, which turned it into what became known as "The Cities' 97", KTCZ.

Unfortunately, it was the ownership which ultimately proved to be the station's worst enemy.

The Union Gets the Boot

One of the biggest hotbeds inside WWTC was the presence of the American Federation of Television and Radio Artists, AFL-CIO (AFTRA).

The AFTRA union was brought into WWTC shortly after Bob Short took over. Short had told staffers of the then news-talk station that because of the station's poor financial situation, news readers would be expected to help solicit advertising along with the sales staff. Outraged, the news people responded by petitioning in the union.

By this time, however, WWTC no longer operated a news department and nobody from the news operation remained but AFTRA was still there, much to Short's dismay. As contract negotiations came up, the Shorts appealed to the staff to petition the union out.

"We're handicapped by the union situation", the senior Short told the staff. "When we're out of the union situation we'll be able to pay you more and have some flexibility in it."

Del Roberts was in the forefront of the campaign to boot the union out. Although Del in fact had helped organize AFTRA at KDWB in the mid-sixties and at WEBC in Duluth a few years later (which cost him his job there), the union didn't seem to be doing much good at this point. The contract specifically outlined that announcers would make just over minimum wage and the union offered no job security or benefits. It no longer seemed worth the dues that everyone was required to pay into it.

There was also a question amongst the staff as to what they would do if a strike was called. Would the disc jockeys be willing to stand outside the WWTC building carrying signs knowing very well none of them were superstars and were considered expendable by the owners?

As Brad Piras remembers, "It was an ugly, ugly scenario. Are we willing to go out, sit out front of WWTC and picket? Who are we? All the people who were there were making minimum wage. None of them had a reputation. It's like the guys at McDonald's; if they decide to go on strike, they get four more guys to make the burgers."

While AFTRA had once been a strong representative of broadcasters, many were beginning to see the union as having more interest in TV news anchors and local actors than in radio announcers. It was described by one veteran announcer as having become "a country club for the Guthrie Theater".

The climax of WWTC's relationship with AFTRA happened at a meeting in the summer of 1980. Dick Driscoll, Del Roberts, Brad Piras and Arne Fogel attended the meeting with AFTRA president John Calin and other union officials.

As the meeting was called to order, Calin demanded that operations manager Driscoll leave, even though Driscoll had been a union member for almost 25 years.

"What do you mean I can't be here", Driscoll protested. I got a card and I pay dues. . ."

"You can't be here because you're management", Calin replied.

"In that case I shouldn't have to pay dues."

"No, you gotta pay the dues."

"Well which is it?!"

Driscoll recalls that Calin lost his cool at that point. "OUT! OUT!", he demanded.

The union officials were also concerned about the presence of Piras since he held the title of program director but he assured them he didn't make any official decisions. As the 'TC staffers and union people debated union verses non-union, Del bluntly asked "Why do we have to be in the union to make minimum wage?" Nobody seemed to have a good answer, and were certainly irritated by the question.

As for the strike question, Calin and the others told them they ought to stand behind the union no matter what, even if it means risking their own jobs, for all union members everywhere.

Arne Fogel, one of the few AFTRA boosters with the station, sat passively through the meeting, until he asked a simple (long forgotten) question.

An AFTRA official pointed at Fogel in a condescending manner, shouting "YOU have no right to be asking that! YOU oughta be smart enough to know better. . ."

"I withdraw the question", a weary Fogel replied.

"What do you mean you withdraw the question?"

The mild-mannered Fogel suddenly saw red. "I withdraw the question because I can't even think of it now because the only thing that's on my mind is to tell you to stop pointing your finger in my face RIGHT NOW or I'll break it off!"

The outburst was extremely out of character for the otherwise polite and laid-back Arne Fogel. The meeting was abruptly adjourned. (The incident did not change Fogel's opinion of the union itself and he remains an AFTRA member today.)

The staff voted the union out of WWTC partly because of Short's promises of more flexible wages and partly because of the AFTRA bureaucracy. As one former 'TC staffer put it, "When the union vote came, I think people were voting more against Calin than the union".

The Shorts succeeded in busting the union at WWTC, but things didn't improve; nobody received any raises and without union representation, the Shorts had much more control over the destiny of the staff — for better or worse.

The Night John Lennon Died

Monday, December 8, 1980 started out as another ordinary day. At 1280 on the dial, it was the same-as-usual madness of the Golden Rock.

But that evening at the ten o'clock hour, the network news came on with a report that would be a shock and devastation to the listeners of this radio station in particular.

"MUTUAL NEWS— Former Beatle John Lennon has reportedly been shot to death outside his apartment building in New York's Manhattan. . ." Reporter Bernard Gershon told listeners what was so far known about the shooting.

At WWTC, the telephone lines were suddenly flooded with calls, some almost frantic, asking if what they just heard was true. Nancy Rosen was on duty that night.

Nancy played Beatles and Lennon music while trying her hardest to handle the situation over the air while taking calls from distraught Beatles fans. Several callers asked to speak to 'TC's resident music scholar, Arne Fogel.

Fogel had just heard the news himself from a friend and was so upset, he had injured his hand after hitting his fist on a stucco wall just before Rosen called him. Rosen asked him to come down to the station and he grabbed a stack of Beatles and Lennon albums from his collection and rushed downtown to the studio.

But when he got there, he found out program director Brad Piras had just called and ordered Nancy not to put Arne on the air — and ordered her not to play any Lennon music or mention the situation on the air.

"Imagine how emotional I was anyway and that made me even more crazy", comments Fogel. He called Piras at home to find out what was going on. Piras, who had been sleeping and had to get up for work in a few hours, wasn't in much of a mood to deal with it.

As Piras remembers, "Arne called me and he's in tears on the phone that Lennon had died and he was extremely emotional. I don't know about out of control but as the conversation went on he did get out of control."

Fogel, for his part, denies he was "in tears" or "out of control" but was certainly upset. He tried passionately to make the point that this was an event that should be covered, especially by this radio station.

"Arne", Piras remembers saying, "you've got to get control. Whatever we do we're going to plan it before we do it. Right now you're too emotional. You won't do a professional job".

"You're fucked in the head!", Fogel snapped back, incensed at being told he wouldn't do a professional job.

Piras made an extra effort to restrain himself. "I'm going to overlook that because you're emotional right now."

Fogel continued to argue the point. Piras told him that not that many people are going to care enough about this for him to make a big deal about it on the air. The conversation finally ended with Fogel slamming down the phone.

Meanwhile, Dick Driscoll tried to call Arne at home to suggest that he should come down and do an on-air tribute. When Fogel's wife told Driscoll he was already at the station, Driscoll called over and asked Arne why he wasn't on the air. Arne told him what Piras had said and Driscoll told him "You hang on, I'll take care of this".

Piras soon called back and reluctantly told Fogel to go ahead and go on, still a little angry about the previous conversation. Fogel thanked him and apologized, then went on just after midnight.

By this time, Steve Bowman was doing his airshift and he and Arne spent about an hour playing cuts from the albums Arne brought in and informally eulogized the slain Beatle. As Bowman did his "dedication line" feature, calls came in from grieving Lennon fans as well as the typical I-wanna-dedicate-this-to-Sue calls. The general mood of the callers that night was somber. As Steve and Arne did their tribute, Mutual cut in with a special update on the shooting.

"Brad really undervalued the impact [of the shooting] and I really don't know why", Fogel says in retrospect. "[But] I was the victor in a sense because how I felt about it turned out to be the way the whole world felt about it."

Looking back, Piras says of Fogel, "Arne is probably the most true music professional that I've ever had the experience to meet. He's probably the most intelligent of the people who worked at the station and I have nothing but respect for Arne. He was mad at me [over the Lennon ordeal] and I understand that. If I'd felt that strongly about something, I'm sure I'd done the same thing."

Throughout the week, Brad and Arne did team up in producing tribute segments to Lennon and a weekend special, which Arne wrote and narrated. But Arne had the feeling that he and Brad were no longer seeing eye-to-eye on things and as for Piras, this was just one small step in his disillusionment with WWTC.

The Man Who Never Returned

When the Golden Rock format began on WWTC in the fall of 1979, the moribund station had very few commercials — Shakey's Pizza Parlor, Insty-Prints and some of Bob Short's businesses were about it.

A year later, it was a completely different story. WWTC had more commercials per hour than almost any other local station. Suddenly, both national and large local advertisers were clambering to buy on WWTC. The station had become a hot spot because of its fast growth, the appealing demographics and the fact that ad time was relatively inexpensive.

A Twin Cities newspaper strike in October, 1980 brought the station even more advertising dollars as ad budgets spent in newspapers were redirected in radio.

The commercial load, however, was doing at least as much harm as good for the station. During the morning and afternoon drive times, when Brad Piras and B.J. Crocker were on respectively, there were up to 20 spots per hour, twice the industry standard. Along with the ads, sports commentator Bob Casey called in during the morning and afternoon shows for live sports reports that could last well over the allotted five minutes. Also, during Piras' morning show, the station sold a ten-minute daily spot to fitness guru Lana Mosow to promote her health spas. Listeners were beginning to complain about all the commercials and research was beginning to show the audience leveling off.

"All the sudden, rather than doing the normal seven commercials an hour, we're doing 20 commercials an hour", recalls Piras. "I got Bob Casey doing sports reports; I enjoyed his shows but Bob Casey talking for five or ten minutes about Bob Short's old cronies did not necessarily fit the image we were trying to project. I had Lana Mosow the exercise lady doing exercises on my show. We went from 90% music down to 60% music."

When the Fall 1980 ratings came out, WWTC showed a slight decline, from 3.9 percent to 3.5. The station's goal was to continue climbing, hopefully into the four and five shares, and beyond. A decline, of any kind, wasn't suppose to happen. Management began to get nervous.

Research was also indicating that listeners were tuning out, not only because of the commercial load, but because too many songs were being played often enough to lose appeal. A song that is fun to hear for the first time in twenty years can get monotonous when it is heard every day.

The Shorts, Charlie Loufek and others saw the problems with listener complaints and negative research as well as Piras did but they weren't about to admit their precious profit-making commercials had anything to do with it. Piras was the program director, so it was his fault, as far as they were concerned.

Piras recalls the situation well. "When things start to look a little shaky, do you step up and say 'Hey it's my fault'? Nobody does that. They're saying 'you're the program director, what the hell is this?'

"I said guys, you're burning the shit out of the music, you're selling dollar-for-hour spots with the newspapers on strike, we're doing everything but playing music and having fun. . .I could just hear the radios clicking off."

Station management looked at the situation from a different vantage point than Piras. Management was under fire from the owners to turn over more profit so the ownership could invest in an FM station and other things. If listeners were turning off WWTC, then the disc jockeys needed to cut their creative bits; cutting the commercial load was out of the question. Piras, on the other hand, looked at the situation more from the listeners' angle rather than a business angle.

Pressured by management to remedy what appeared to be a potential loss of audience, Piras made a proposal to the owners and management team in February, 1981. Piras proposed cutting the commercial load in half, dropping Bob Casey, Lana Mosow and some of the network features, and resting some of the over-played oldies by mixing in either more contemporary hits or even country-rock by such artists as Charlie Daniels and Hank Williams, Jr. Country-rock would mix well with the rockabilly the station played, Piras contended, and would do well to attract the target demographics.

After giving the proposal his all, the managers and owners flatly told him no.

"Oh by the way, Brad", Piras recalls being told by general manager Loufek, "your show, way too much talk. We're looking at your numbers.

If you can't turn that thing around, we're going to have to put Del or someone there."

"Oh you would?" Piras responded. "So you don't want to do anything about the music, you won't do anything about the commercials, Bob Casey's gonna stay?"

"Yep."

"Lana Mosow's gonna stay?"

"Yep."

"You're gonna start having people cutting back on their bits, is that right?"

"Yep."

"And you're looking at my numbers, they're starting to slide so there's the potential you'll put Del in, is that right?"

"Well, we probably won't go that far. . ."

Days after the meeting, Piras was assigned to attend the annual Buddy Holly memorial concert in Clear Lake, Iowa and do remote broadcasts for WWTC. He loaded all of his belongings in his car, drove down to Clear Lake and made his final on-air appearance for WWTC. He drove on to Portland, Oregon and took a job at a radio station there and like Scott Carpenter before him, he didn't look back or say goodbye.

Del Roberts was slipped into the morning shift and was given the title of "acting program director". Shortly before Piras left, Arne Fogel was given the newly-created title of music director. Since Piras' departure came so soon after Fogel acquired that position, Fogel was convinced for a long time that it was his fault that Piras had left, especially in light of their dispute over the death of John Lennon. (Piras says that he would have given Fogel full support as music director had he stuck around.)

If Del Roberts and Arne Fogel ever had any kind of a friendship, it certainly hit the rocks at this point. The two became thorns in each other's side.

While Arne was expanding the music library, Del felt the selections he was adding were too much along the line of what he considered to be "acid generation" hits, which he felt were inappropriate for the format. To Fogel's chagrin, Del pulled the tapes of music Arne had added and brought them into manager Loufek's office.

"He's into the art of the music", comments Roberts, "and I'm into the success of the music. He played a record because he felt it was important in the career of an artist, I played it because it got ratings."

"That's pure bullshit", retorts Fogel. "I was fulfilling what was the destiny of what was started. Del wanted to take it in a different direction; more fifties and sixties. Del Roberts has a very distorted view of what pop music is all about."

Del felt the same way about Arne's view of pop music and the friction between the two increased.

Meanwhile, the staff of WWTC was preparing to move to a new studio and facilitics.

The Move to Wesley

At the time Bob Short took over WWTC three years earlier, the Minneapolis papers reported he intended on moving the station to his Wesley Temple Building, located a few blocks away from 'TC's then-current location at the Builders' Exchange building in downtown Minneapolis. The station was paying $19,400 in rent annually, a situation the thrifty Shorts didn't care for, especially when they owned several downtown buildings.

The Wesley Temple Building was a historic landmark which sat next to the Wesley Church on East Grant Street. A large neon sign atop the building flashed the message "Worship at Wesley Church".

The Shorts were stuck paying rent for space in the old building on Second Avenue a lot longer than they would have liked. There were still a few more years left on the lease when they took over and were unable to break it. They also had some grandiose plans for the new facilities that they couldn't quite work out.

"It was delayed for several months because we couldn't figure out if we were going to be on the 11th floor or the 12th floor", recalls Driscoll. "The reason there was a big debate on that was because Short wanted his radio station to be in the middle of a restaurant that he was going to put up there."

The idea was that the restaurant would be located on one floor with the control booth of the radio station right in the restaurant, and the rest of the radio facilities on the other floor.

"The city [inspector] and the fire marshal took a look at the Wesley Temple Building and said the only way you're going to have a restaurant up here is if you widen these stairs, put in an entirely new sprinkler system and whatnot", recalls Driscoll. "That took care of that idea".

It was decided that the radio station would be on the top floor of the twelve-story building. Ironically, this was the exact location of the old WTCN Radio in the 1930s and '40s. WWTC was, in fact, the former WTCN Radio.

Del Roberts gave his mother a tour of the new facilities once the station moved in. She told of how she went up there as a girl to see the big bands play for a standing-room-only crowd live on the air on the old WTCN. She pointed out where the bandstand was, where the spectators stood and where the young people danced.

The move from Second Avenue to the Wesley Temple Building took place over a three month period. The station was completely moved in by May, 1981 but the facilities were moved piece-by-piece. Instead of hiring a professional moving company, a Ryder truck was rented and the 'TC staff itself did the moving, with some staffers helping out by loading stuff into their own station wagons and vans as well as the rented truck.

Mike Ryan recalls, "I remember one time I was late for my airshift because Dick Driscoll and I were up in the new building hauling stuff in and I completely lost track of time. We were loading up anything we could move, anything we could carry, and we were running back and forth, going up the elevator and I was suppose to be on the air at noon at the old place. I completely lost track of time and I think I got back at like 12:30. Sorry I'm late!"

Eventually everything from the old building was moved out but the one thing that was left behind was the red and white two-panel sign that said 'WWTC' vertically and was displayed on the outside of the old building.

The sign, which had been there since the mid-sixties was not owned but merely rented from the sign manufacturer. Driscoll recalls the sign company wanted to sell it to the station for a token amount of money but Bob Short refused to but it. The sign company was stuck with a gaudy sign that was of no use to anyone else.

Along with the sign, rooms specifically designed to be a radio studio were left behind. There had been radio facilities there since the fifties when WDGY occupied the floor. WWTC moved into rooms that too were once a radio studio, but it had been some forty years.

Says Arne Fogel of the Wesley studios, "It looked like someone took a suite of lawyers offfices and put a sign on the door that said 'radio station'."

Radio 1280 might have broadcast there a long time before, but the spirit of what was known as the "Radio Station of the Twin Cities Newspapers" had long since evacuated.

THE GOLDEN ROCK

OF MINNEAPOLIS/ST. PAUL

HARRY ZIMMERMAN
FRANK BUETEL
SEV WIDMAN
PAUL SEVAREID
LARRY FISCHER
JACK THAYER
STU MANN
SKIPPER DARL
Your Favorite Stars are on
WTCN
RADIO and TV
Dial 1280
Channel 11
ABC NETWORK

2. STARBURST CLUSTERS

From the Remnants of WTCN

WWTC-AM 1280 is the sixth oldest radio station in Minnesota. Founded by Dr. Troy Miller, the station took to the airwaves on August 10, 1925, with the call-letters WRHM, for "Welcome Rosedale Hospital Minneapolis". WRHM broadcast from the hospital, located at 4429 Nicollet Avenue with only fifty watts of power.

When the *Minneapolis Tribune* and the *St. Paul Pioneer Press-Dispatch* jointly purchased WRHM in 1934, the call-letters were changed to WTCN, for "Twin Cities Newspapers". The station was initially located at 1250 on the dial.

WTCN radio broadcast from studios on the 12th floor of the Wesley Temple Building at 123 East Grant Street in downtown Minneapolis — from the same rooms that became the studio and offices of WWTC in the early 1980s — throughout the 1930s and 40s. The station was affiliated with the NBC-Blue network. At the time, NBC operated two separate radio networks, the main one was called NBC-Red, the second one NBC-Blue.

Because of an international treaty known as the North American Regional Broadcasting Agreement, 795 radio stations in the United States assigned above 730 kilocycles had to be reassigned new frequencies effective March 29, 1941. The treaty affected all seven of the radio stations on the air in the Twin Cities and on that date, WTCN moved from 1250 to 1280 on the dial.

When anti-trust legislation forced NBC to sell its second network in 1943, the Blue network was sold to Edward J. Noble of the Lifesavers Candy Company and was renamed the American Broadcasting Company. WTCN remained a corner-stone ABC affiliate. Early ABC programs featured the likes of the Lone Ranger, Bing Crosby and Walter Winchell.

In 1949, WTCN was granted a television license. The TV facilities were set up in the Radio City Theater building at 50 South Ninth Street in Minneapolis and the radio station moved there as well, WTCN-TV went on the air on July 9, 1949 on channel 4.

Three years later, WCCO purchased channel 4 from WTCN. Having the desire to stay in the television business, WTCN applied to the Federal Communications Commission for channel 11. While WCCO took over WTCN's Ninth Street studio, WTCN moved its facilities to the Calhoun Beach Club in south Minneapolis.

But St. Paul radio station WMIN had also applied for channel 11. The two competing radio stations worked out an agreement an a joint application was submitted.

On September 1, 1953 both WTCN-TV and WMIN-TV began operating on channel 11, the two stations alternating in hour and a half intervals throughout the broadcast day. The station was affiliated with the ABC Television Network. In 1955, WMIN sold its share of the TV station to WTCN and Channel 11 became, wholly, WTCN-TV.

WTCN radio and television continued to broadcast jointly from the Calhoun Beach Club as the local ABC station into the early 1960s.

WTCN Radio carried the ABC features along with local talk and music programs. But the radio industry was rapidly changing. Television was quickly taking over as the dominate home entertainment and network/variety radio stations were eating dust in the process.

Some stations, such as WDGY and KDWB were surviving well playing rock 'n' roll. Other stations were programming other types of music.

ABC switched affiliation to KMSP-TV, Channel 9 in 1961 and KRSI Radio, 950 in 1963. While WTCN-TV replaced network programming with reruns, movies and sports, WTCN Radio filled the void with a structured "beautiful music" format.

Finally in 1964, WTCN's parent company, Time-Life, sold the stations to new owners. Chris-Craft Industries, the company that bought the TV station, was convinced the radio industry was dead and wanted nothing to do with the radio station. The radio station was sold to Buckley-Jaeger Broadcasting of New York for $5,000 with Richard D. Buckley holding approximately 85% interest, John B. Jaeger and Richard Buckley, Jr. holding the rest. While the television station retained the call-letters WTCN, the radio station signed on as WWTC on October 2, 1964. Vice-president and general manager of WWTC was Robert V. Whitney, succeeded by Richard J. Korsen a year later. The program director was Bobby Oaks.

Beautiful Music For the Twin Cities

Tuning in 1280 on the radio in the fall of 1964, the listener was treated to light classical and easy listening music with specific selections played during each part of the day to set a specific mood.

Emphasis was away from the personalities of the announcers; the announcers were not even allowed to say their names on the air. Instead, each part of the broadcast day had a pseudo-romantic title to convey the "mood".

The station signed on at six in the morning with what was called "A.M. Overture". At nine o'clock, there was "Serenade". As the day went on, other "moods" included "Limelight", "Carousel", "Gaslight", with the evening coming to a close at midnight with "Quiet Hours".

After a selection of music, the announcer told the listener what selections had been played over recorded harp music ("a scratchy harp", recalls morning announcer Dick Driscoll).

From eight to nine o'clock in the evening, "Conductor's Choice", with host and commentator Dr. Frederick Fennel brought listeners the finest in classical music performances and Broadway musical selections conducted by the likes of Arthur Fiedler and Leonard Bernstein. "Conductor's Choice" was sponsored by Twin City Federal bank.

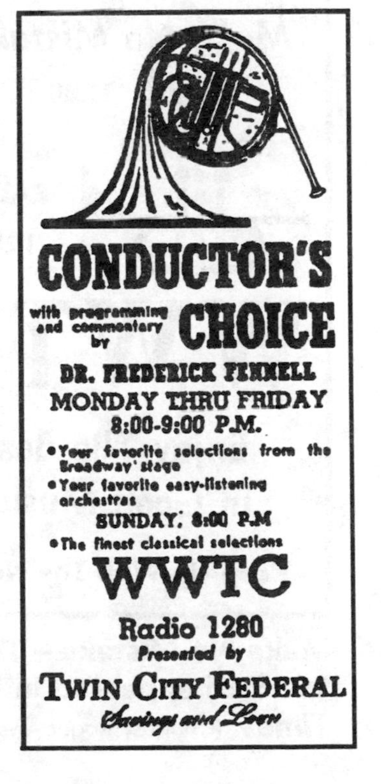

This format had been running on WTCN for a year and a half and continued for just over two months after the station became WWTC. In mid-December, 1964, program director Bobby Oaks phased WWTC into a slightly more progressive format, shifting the emphasis away from classical music and the station began playing the likes of Ray Conniff, the Living Strings and Bert Kaempfert.

Announcers Loyal Farrel, Tom Wynn and Jack Houston were finally allowed to say their names on the air. The station at this point began a gimmick called the "starburst cluster". A "starburst cluster" was three songs played back-to-back without commercial interruption. But as *Minneapolis Star* columnist Forrest Powers observed, "After each record...the station interrupts the music to announce there will be no commercials."

WWTC was on such a low operating budget that announcers were doing split shifts. Loyal Farrel was on from 5:30 AM to 9 AM and from 11 AM to 1:30 PM. Tom Wynn was on from 9 AM to 11 AM and from 1:30 to 6 PM, with Jack Houston from 6 to midnight.

Dick Driscoll, who had been an announcer and personality at WTCN Radio and Television and continued at both stations after the ownership separation, decided he had enough when asked to do split shifts. The former WDGY rock jock disliked the dull, non-personality format enough, but this was too much.

Make No Mistake – This 1966 ad appeared in *TV Times*. (Courtesy Roger Awsumb)

As Driscoll remembers, "Bob Oaks...came in from the east to run this thing and he straight-jacketed us as personalities. I told him I thought I should look for greater pastures and I did, I went to KQRS." (KQRS was not a rock station at the time.)

Tom Wynn and Jack Houston soon left as well, replaced by Dick Haase, Chuck Englund and former KDWB DJ Lou Riegert (Riegert today is known to Cable News Network viewers as Lou Waters).

WWTC's sales manager was a young man by the name of Peter Lund. At age 23, Lund came to WWTC after working as a page at WCCO while still in high school, and graduating from the University of St. Thomas in

St. Paul. He left the station in 1966 to become the general sales manager for the Westinghouse Broadcasting (Group W) stations, later moving on to other things and eventually coming to New York where he joined the CBS Television Network in 1977. In February 1995, the former WWTC salesman was appointed president of the CBS Broadcast Group.

Although WWTC had a limited budget, program director Bobby Oaks had a taste for glamour and glitz, or at least the illusion of. To show a supposed commitment to news, WWTC purchased a fleet of 35 news cars even though the station didn't even have 35 employees. Oaks got WWTC a flattering profile in a book published by the Minneapolis Chamber of Commerce, which described the station as having "the exciting flavor of show business, with news, features and America's favorite music. WWTC stands for Wonderful World of the Twin Cities, where it has found early and gratifying success." The success, if any at the time, wasn't that gratifying, but at least the publicity was good.

When the radio station was sold separately from WTCN-TV, it was forced to move out of the Calhoun Beach Club studio — after spending some $5,000 remodeling and putting the finishing touches on the radio studio.

Luckily, WDGY was abandoning is studio on the second floor of the Builders Exchange Building on Second Avenue in downtown Minneapolis. Left behind was a ready-made radio studio for WWTC, a unit that would have been useless to any other tenant with the extra-thick walls, indoor windows and broadcasting equipment.

The station adopted the letters WWTC because as a separate entity from WTCN Television, federal regulations required the name change. WWTC sounded close enough to WTCN for the station to somewhat keep its former identity. While the letters were said to stand for "Wonderful World of the Twin Cities", a sign displayed across the windows on the building read "World Wide Twin Cities News".

WWTC limped along through the sixties playing sweet music in "starburst clusters". In the American Research Bureau (now known as Arbitron) ratings, WCCO (830) had some 50 percent of local radio listeners. What was left over primarily went to top-40 stations WDGY (1130) and KDWB (630). KSTP (1500) was doing better than WWTC in the ratings playing the same type of "easy-listening" music.

Adding insult to injury, the owners of WWTC didn't seem all that interested in that station's success; any profit the station did make was drained by parent company Buckley-Jaeger to purchase the more prestigious WIBG, a 50,000-watt rock station in Philadelphia.

Ratings mean revenue in the commercial broadcasting business and by April, 1969 when Charles Tyler III replaced Richard Korsen as general manager, WWTC was in deep trouble. Some big changes had to be made and to do it, the station lured back Dick Driscoll.

The New Season of Sound

After successfully turning KQRS-FM (92.5) from a classical music service to a jazz-by-day, "'underground" rock-by-night station, Dick Driscoll promised to turn WWTC around as its new program director.

Driscoll replaced the "starburst clusters" format (which he jokingly calls "peanut butter clusters") with what was promoted as "The New Season of Sound". The music on WWTC became more up tempo and progressive. 'TC listeners were now hearing selections from artists such as Peter, Paul and Mary, the Fifth Dimension and the Youngbloods, plus rather heavy album cuts by the likes of Ray Stevens and Rod McKuen, as well as mild instrumentals and vocals from the likes of Frank Sinatra.

Along with the more progressive music, Driscoll brought in a new staff of talented air personalities. Randy Cook, Joel Larson and Paul Stagg handled the daytime shifts, with Driscoll in the afternoon and Tom Ambrose, the station's music director, in the evening. The disc jockeys were encouraged to loosen up on the air, tell jokes and be more personable to the audience.

On-air slogans on WWTC at the time included "The New Season of Sound" and "Music For Beautiful People". Paul Stagg's young daughter even did a spot for the station, saying in a child's voice, "Music for beautiful people, WWTC; That sounds logical!" There were also smooth and mellow jingles on the station, singing "W-W-T-C/Ra-di-o Twelve-Eight-O".

Beginning in January 1970, WWTC began broadcasting on a 24-hour basis, and the station went on an aggressive advertising blitz, promoting the artists the station played as well as the air personalities.

A decade before the Golden Rock, WWTC promoted its air personalities, including Dick Driscoll, in this 1970 advertisement.

In an advertisement that appeared in the winter, 'TC claimed to be "the winter warmer". The ad featured a transistor radio wearing earmuffs. A summertime ad showed the same transistor radio on ice.

Driscoll's formula proved to be a success. The ratings made a complete turnaround and soon the once-struggling "elevator" music station was holding its own. In a radio market where success was measured only in the shadow of the mighty WCCO, WWTC was competing quite well, especially in the adult male demographics. In overall ratings, WWTC ranked in fourth place, behind number one 'CCO and neck-in-neck KDWB and WDGY, and among 25-49 year-old males, it was in second place. WWTC was becoming a force to be reckoned with.

The element of entertainment was a strong part of 'TC's format. During his afternoon show, Dick Driscoll would often portray several different alter egos, with names like "Flying Officer Nelson", whose traffic reports were always screwed up and "Maintenance Man", who would cause on-air "technical difficulties". (These alter egos returned to WWTC ten years later with the 'Golden Rock'.)

In 1971, Glen "Big Daddy" Olson was hired to do WWTC's midday shift. Olson had actually worked at WWTC a few years before, on the air as "G. Edward Foshay" during the 'beautiful music' period. He would often parody TV preachers such as Billy Graham and Oral Roberts on his radio show.

JOEL LARSON WWTC 1280 Radio

(Courtesy Art Phillips)

Other air personalities hired at about this time included Greg Gears, Jim Teeson, David Teller and a young man named John Hines.

Hines turned up on WWTC doing weekend and fill-in shifts. He would later became a well-known morning personality on KSTP-AM, WLOL-FM and eventually, KEEY-FM (K-102). He told the *Minneapolis Tribune* in 1978 (after becoming a star on KSTP with partner Charlie Bush) that he was fired twice by WWTC "once for economic reasons, then I was hired back under a different budget and fired again for economic reasons".

The most popular — and controversial — personality on WWTC at the time had to be morning man "Fast Eddie".

Fast Eddie In the Morning

Paul "Fast Eddie" Lowell

Paul "Fast Eddie" Lowell had a style that would have fit in well ten years later with WWTC's "Golden Rock" format. His on-air stunts were hilarious, unpredictable and occasionally risqué.

Originally from Illinois, Eddie was a talented musician, playing piano in bars and with jazz bands. He came to Minneapolis in the 1960s and learned broadcasting at Brown Institute.

According to Dick Driscoll, Fast Eddie was one of the reasons the Federal Communications Commission instituted rules about telling the other party on the telephone beforehand that they are on the air. He was known to do such things as call a pizza parlor on the air and say "we need 500 pizzas, we'll be down there in five minutes to pick them up. You think you can handle it?" He would tell the bewildered pizza baker later who he really was.

Fast Eddie provided a "wake-up call" service, calling people and telling them he was going to be stopping by, and then sometimes even showing up in person, nude and wrapped in cellophane.

Often Driscoll himself was Fast Eddie's accomplice in his on-air pranks. In the tradition of Alan Funt, Dick and Eddie went out and did "candid microphone" bits for 'TC.

One such prank had Eddie going into the Dayton's store in downtown Minneapolis with his hidden microphone and telling a saleswoman he was interested in buying a bathing suit for his girlfriend. When asked the size, he told the clerk he didn't know the size but he and his girlfriend are about the same size, so "could I try it on and see how it fits?"

As the perplexed sales clerk was about to call security, Driscoll stepped in to the rescue.

"Hi, you've been talking to Fast Eddie of WWTC Radio 1280, this is all on tape." Over a good laugh, Dick and Eddie had the sales clerk sign a release and gave her a couple of free radios, telling her to "tune in tomorrow to the Fast Eddie show". The two disc jockeys then left the store quickly while they still had their skin.

In another situation, Eddie brought the candid microphone to a Northwest Airlines ticket office, asking how much it would cost to fly to Los Angeles. After being told, he asked "how much in freight", telling the airline employee he intended to ship himself to Raquel Welch in a box. Sometime before, he called the actress on the phone, on the air. She immediately hung up on him.

There was also the time Driscoll pulled his car into a Standard service station with Fast Eddie in the back seat, once again with the candid microphone. Driscoll told the attendant to "wash the windshields, check the oil, fill it up with gas — if the guy in the back says anything to you, just play along with him because he's crazy. We're taking him down to St. Peter and he doesn't know it so just humor him." Driscoll then walked off while Eddie made nonsensical remarks such as "It wouldn't be so hot if it wasn't so cold", with the gas station attendant just a little uneasy about the "crazy man" in the back seat.

There were those who thought Fast Eddie went too far. A show he did on the morning of October 22, 1971 prompted a complaint to the Federal Communications Commission. This show featured a telephone caller who claimed to be a male hairdresser. There were subtle references to homosexuality in the bit, all intended as good, clean fun.

But University of Minnesota student body president Jack Baker wasn't amused at all. Baker, a gay activist, was outraged because the "male hairdresser" made references to him in the on-air bit.

Baker, along with two real-life male hairdressers, filed a complaint with the FCC asking the agency to order WWTC to give them airtime to reply to what they deemed as slurs against them.

These ads promoted WWTC's " New Season of Sound" in *Twin Citian* magazine in 1970.

The official complaint claimed the on-air phone caller "discussed same-sex marriages in a manner designed to bring ridicule and loss of respect to complainant Baker" (Baker at the time was lobbying to get same-sex marriages legalized in the state of Minnesota). The complaint also said the context of the caller's statement "constitutes an attack on the honesty, character, integrity and like personal qualities of hairdressers as a group."

Baker and the two hairdressers went on to allege the caller was not a bona fide member of the listening audience but either Fast Eddie himself or another WWTC staff member. If that wasn't bad enough, Jack Reynolds, who became general manager of WWTC in May 1971, refused to apologize for the matter, according to Baker.

Reynolds contended that Baker, being a known public figure, "falls into the basic classification category as the President or Vice President as the butt of jokes."

The FCC's rules for equal time considerations for opposing views applied only to political views and commentaries, not entertainment. Baker's request was denied and the case was dropped.

The Live Concert Series

A periodic feature of WWTC in the early 1970s was live concerts. The station would broadcast a performing artist live from the WWTC studio or from Sound 80 recording studio in Minneapolis.

Beforehand, a station personality would interview the artist or group. The interviews were often quite interesting.

When Tom Ambrose interviewed the Fifth Dimension, Dick Driscoll coached him ahead of time, telling him not to ask such mundane questions as "how long are you gonna be in town", "where are you going next", "how did you get into music", etc. Driscoll suggested to Ambrose that he open the interview by asking out of the clear blue, "You've just made an album called 'Aquarius'; do you believe in the signs and if so, what sign are you?" Having been a relatively new question at the time and certainly something they had never been asked before by an interviewer, the Fifth Dimension spent a full hour with Ambrose rather than the few minutes they would have given any other radio interviewer.

Another memorable concert was performed by the Doobie Brothers in the fall of '71. Although the Doobies were more of a rock band and hadn't yet become popular (their first hit single came out the next year) WWTC aired the concert, which consisted of an acoustic set and an electric set, live from Sound 80. The show received a warm reception and helped WWTC expand its audience into younger demographics.

Driscoll did the before-hand interview, embarrassing himself just slightly when he asked "None of you have the name 'Doobie' and none of you are brothers, why do you call yourselves 'Doobie Brothers'?" Being ignorant that "doobie" was a common slang for marijuana, Driscoll recalls members of the soon-to-be supergroup looking at him as if he were "really square".

The Doobie Brothers concert on WWTC was reportedly the first live rock show aired locally by a Twin Cities radio station. While it would seem likely the show would have aired on a progressive FM rock station such as KQRS, there was less of a chance the show would be recorded and "boot-legged" airing on the AM-mono WWTC.

Good Morning, Athletic Supporters

Appearing with Fast Eddie in the morning beginning in January 1972 was sports guy Dave Sheehan. Sheehan started out doing one five-minute sports report on WWTC every morning, Monday-Friday. His duty was later expanded to five reports a day, five days a week.

Sheehan began his reports by saying "Good morning, athletic supporters, this is Jock Talk with Dave Sheehan". He then proceeded into an outspoken, arrogant and often amusing view of the world of sports; the kind of guy any sports nut loved to hate.

He once told *Minneapolis Tribune* columnist Robert T. Smith his three personal heroes were himself, Sid Hartman (sportswriter for the *Tribune*) and Howard Cosell (of ABC Sports) and that "my first purpose is to entertain and then inform. I appeal to those with a higher level of intelligence, the 'with-it' people."

Calling himself "the mouth you love to hate five days a week", Sheehan's remarks generated laughs and anger.

"I very nearly missed the Twins on television", he once told his listeners. "I saw 'The Curse of the Undead' on Channel 11 and watched it an hour before I realized it wasn't the Twins."

In another broadcast he quipped, "I see the University of Minnesota-Morris lost their head coach. I hope they find him before the season starts."

He suggested once that after seven dull Super Bowl games that it should simply be called "The Last Game of the Year", and he once spoke of someone he knew who was so naive he thought "oral sex was talking about it".

Al Flom, then the assistant engineer at WWTC, remembers Sheehan well. "He was a wild man. He was the Sam Malone of WWTC. I remember Sheehan had this regular girlfriend who was built just like you wouldn't believe. Beautiful girl. And we had the control room and off to the right there was a window and a small booth. That's where he'd do his sports."

"One time he was doing his sports and his girlfriend thought she'd try to distract him. So she did a striptease right in the control room — all the way down. Greg Gears was on the air and he was just [stunned]. It was unbelievable."

Love him or hate him, WWTC was willing to be his "athletic supporter"; his flippant sports updates drew a listenership of up to 50,000 in the early seventies, more than that of a lot of the station's music programs.

The Experimental Program

In a one-week venture in May 1972, WWTC aired a strange series of documentary segments called "Perception".

The creation of a young newsman named Rick Leepart and producer Mike McKenzie (who was the chief engineer at WWTC with Al Flom as assistant), the three-minute segments aired at 7:30, 8:30 and 9:30 PM for a week and covered a wide variety of subjects.

Some of the segments dealt with the adventuresome Leepart's personal experiences posing as an inmate at Stillwater state prison, skydiving (he brought a tape recorder with him as he jumped out of a plane), and being hypnotized on microphone, revealing to awed listeners that in a previous incarnate he was an English gentleman of another century.

In other segments, Leepart interviewed interesting people including a Hennepin Avenue prostitute, whom he took out to dinner, getting her view-points of the world, and Archbishop John Roach, whom he asked what a Catholic Bishop does about sexual desires.

After WWTC's initial airing of "Perception", the program was offered to other stations in other cities via syndication.

WWTC Spectrum Radio

In June 1972, sales director Lee Zanin was promoted to WWTC general manager after Jack Reynolds left to join sister station KOL Radio in Seattle. When Zanin took the position, the station was holding up well in the ratings and he kept things going without any major changes.

Then in early 1973, when the ratings came out, Zanin was horrified to see a supposed loss of almost 75 percent of the audience from the previous ratings period. 'TC dropped from fifth to twelfth place in one book.

Zanin's first suspicion was that the book was a "fluke". So he went to the headquarters of the American Research Bureau to inspect the figures for himself. Although he found evidence the plunge in listenership might not have been as bad as reported, there was still a significant drop.

WWTC's target audience was in the "middle demographics", the 25-49 age group. Zanin took a phone survey and came to the conclusion that the progressive sound and hip disc jockeys were "too up tempo for the age group we want to cover". Live rock concerts and surrealistic documentaries were a little too much for the pre-hippie generation.

The first thing to go was the wild and crazy morning man Fast Eddie. While his buddy Dick Driscoll had resigned as program director a few months before, new program director Paul Stagg didn't care much for Eddie, explaining his dismissal by saying, simply, "we agreed to disagree".

A few weeks after leaving, Fast Eddie turned up as the afternoon jock on WLOL-AM (1330) and later went to KSTP-AM (1500). But as the business of radio changed, gigs for personality jocks such as Eddie became more scarce and he wound up spending his last decade doing janitorial work. He succumbed to cancer on December 28, 1995, at the age of 60.

Driscoll continued as a weekend announcer and station engineer. He wanted to concentrate on the technical side of radio and pursue outside interests.

Glen "Big Daddy" Olson, who had done the midday shift, replaced Fast Eddie. Olson had a more folksy style. While on the air, he imagined himself talking one-on-one with a guy in his middle thirties, sitting across from him.

As he described his style in a *Minneapolis Star* profile, "I really hate guys who come on the air and say 'Hi everybody'. That's fake. It's impersonal. It's bad. So I figure there's one guy there and I'm talking to him and we're having a session." He said he gives his imaginary friend "a few thoughts while sitting, standing or shaving".

With Olson in place of the somewhat crazier Fast Eddie, Zanin restructured the WWTC format to what was called "WWTC Spectrum Radio". "Spectrum Radio" was, in the words of Zanin, "the full spectrum of information and entertainment".

The new "spectrum" format included a mellower musical sound, structured to consist of an older middle-of-the-road hit (Frank Sinatra, Bing Crosby), followed by a cover instrumental (elevator music), followed by a top-40 "easy-listening" tune (Roberta Flack, the Carpenters), followed by commercials, etc.

(Courtesy Art Phillips)

In between music sets and ads were so-called "information segments" every 20 minutes, which included news, the infamous Dave Sheehan sports reports, and Linda Morgan on "women's issues". Zanin assembled what he claimed to be the largest radio news staff in town, which included Tom Wayne as news director with

Dave Hoglin and Rick Wagner as news readers, along with reporters and other news staff.

Fireworks on the Fourth

WWTC took pride in being part of the community, participating in charitable events and doing live remote broadcasts from places all around the Twin Cities.

In February 1974, the station raised $383,000 in a week-long radiothon for leukemia. WWTC broadcast live every year from the Minnesota State Fair, the St. Paul Winter Carnival, the Minneapolis Aquatennial, Duff's Golf Tournament and other local events.

One such event, however, proved to be a frightening experience. On a steamy July 4, 1974, WWTC set up the camper used for remote broadcasts at Lake Calhoun, in the Thomas beach area on the south-west shore. The camper, graced with banners reading "WWTC SPECTRUM RADIO 1280", was equipped with turntables, a microphone and a sound board. Evening announcer David Teller, along with engineer Al Flom, broadcast from the beach in air conditioned comfort as hundreds gathered around the shore, the sun slowly setting, gearing up to observe a fireworks display. Families with small children, elderly couples and groups of young people from all over Minneapolis showed up, some waving at Teller through the window on the WWTC camper.

But certain members of the crowd, some of whom were drinking heavily, began to get unruly early on. Firecrackers were being set off hours before the official display began.

By ten p.m., shortly after the grand finale, a number of youths began lighting their own fireworks and throwing them indiscriminately into the crowd. Some people, especially those with small children, began to panic and run, while others joined in the melee, throwing rocks and bottles. An all-out riot broke out at Lake Calhoun. Windows in parked cars were smashed and some even threw objects at passing traffic.

"I was amazed there weren't any injuries from firecrackers", a park policeman told the *Minneapolis Star*. "I saw one little kid almost get hit. People were just throwing fireworks up into the air."

In the middle of it all was the WWTC Spectrum Radio camper. Al Flom recalls, "People were coming up and starting to rock the thing and the needle was skipping all over the record…it was pretty scary for a while."

When a couple of youths shot bottle rockets at park police, Minneapolis police were summoned and they charged the crowd in riot gear, dispersing everyone with dogs and teargas.

Three young men and one juvenile were arrested on a variety of charges, including disorderly conduct, inciting to riot, criminal damage to property and breach of peace. No serious injuries were reported, including Teller and Flom, who, locked inside the camper, were in probably the safest place they could be.

End of the Spectrum

As "spectrum radio", WWTC continued to limp along with mediocre ratings, never quite getting back up to the success it achieved at the turn of the decade. At the end of 1974, Dave Sheehan left WWTC to become sports commentator at WRC-TV in Washington, D.C. Sheehan had been WWTC's most popular personality, the only one really keeping the station afloat. (A few years later, he returned to the Twin Cities, stirring up ire as the sports commentator on KMSP-TV, Channel 9).

With Sheehan's departure, WWTC struggled even harder to keep above water. Lee Zanin knew things couldn't continue as they were and strongly considered the avenue of all-out format change.

"...and they call us pigs."

WWTC News has been talking to cops for the past three months. Specifically, members of the Minneapolis Police Department.

Do wives worry about policemen husbands? Are cops prejudiced? How do cops view the local court system. Is there police brutality?

These are some of the questions we asked the rookies, the veterans, the cop who stopped you last week for speeding, or helped you last month.

And we got the answers. Now you supply the conclusions. Don't miss our one hour special. It could change your views.

A WWTC News Special Presentation
Sunday, November 22 at 4 p.m. WWTC Radio 1280.

CITIAN

News had been an important element at WWTC. It would soon be even more important to the station.

Today WWTC radio trades violins for violence.

June 18, WWTC radio replaces every bit of music with news. Now, we're not trying to put down all-news radio stations.

But if you listen to radio to soothe your nerves, reports of mideast tension, recession and crime just might produce the opposite effect.

That's why we'd like to call your attention to the music of KEEY-AM at 1400 on your standard radio dial.

The best of what you're about to give up on WWTC is what we play all the time on KEEY. Along with enough news to give you information, but not anxiety.

What's more, we do it selectively enough to fit just the mood you want. At the time of day you want it.

1400 AM

Think of it this way. You're not losing a WWTC, you're gaining a KEEY.

3. TRADING VIOLINS FOR VIOLENCE

All News and Nothing But

Lee Zanin was a bit of a newshound. As general manager of WWTC-1280, he boasted of having the Twin Cities largest radio news staff. He considered news to be an important part of a radio operation.

As WWTC struggled in the ratings as a middle-of-the-road music station in late 1974, Zanin strongly considered turning WWTC into an all-news operation. But while all-news stations had proven success in many cities, the costs of building and maintaining one were too prohibitive for Buckley Broadcasting, the owner of WWTC.

Then in 1975, NBC radio announced it would be offering a 24-hour news service to stations called the News and Information Service Network, or NIS. The NIS service was to be a separate operation from the NBC Radio Network.

NBC radio chief Jack Thayer, a Chicago native who was raised in Minnesota and had worked for several area stations including WDGY, WLOL and WTCN Radio and Television, conceived the whole operation. NIS was to offer up to 50 minutes an hour of national and world news, packaged in a way that would allow stations to carry as much or as little of the service as they wanted by keeping the segments short. The remaining ten minutes in the hour that NIS kept open was to be used by the station for local news.

Zanin was quickly sold on the concept and WWTC was one of the first stations to sign up with NIS. The arrangement WWTC made was to carry 30 minutes of the network feed each hour (broken into 15-minute segments), the remaining time to be used for local news and features.

On Monday, June 18, 1975, the turntables were shut off for good at WWTC and the all-news format began on the same day the NIS service started up. The entire disc jockey staff was dismissed except for Dick Driscoll, whom WWTC hung on to because management thought he had a good voice for news reading and announcer Linda Morgan, commenting on "women's issues".

As news director, the station hired Ken Trimble from Philadelphia. Having worked at some of the big all-news stations on the east coast, he brought with him an east coast temperament which came out especially when dealing with the young, inexperienced "hicks" at WWTC (most of whom were in their 20s). He was known for raising his voice and for throwing things around the newsroom.

Driscoll soon grew tired of just reading news and went on to pursue a project with the Corporation for Public Broadcasting along with former 'TC disc jockey Tom Ambrose. Ambrose, along with former 'TC jocks Paul Stagg, Greg Gears, Joel Larson and Jim Teesen went to WCCO-FM (102.9), which at the time was doing a personality-oriented middle-of-the-road format similar to that on 'TC, only in stereo. Glen "Big Daddy" Olson began doing mornings on WLOL-AM (1330), later moving up to the big 8-3-0, WCCO. Driscoll, meanwhile, found work announcing for both "beautiful" music station WAYL-FM (93.7) and WTCN-TV, Channel 11.

Listeners who turned to 1280 AM on June 18 heard a smorgasbord of network reports handled by two-man and man-and-woman anchor teams,

(Courtesy Art Phillips)

plus other network features including book and movie reviews, analysis and commentary and other newsworthy departments. The network segments began with an anchor saying "This is your News and Information Service."

It was almost a precursor to cable TV's CNN Headline News. The local segments were handled by newsreaders including Driscoll and five others, who read mostly wire-service material, and a couple of stringer reporters.

Under news director Ken Trimble, those hired to report news at WWTC included Tom Myhre (who once did WDGY "20-20 News" under the name Tom Mack), Bob Berglund, whose later-to-be morning partner on WLOL-FM, John Hines, had been a disc jockey at the old 'TC and Paula Schroeder, one of the few women reporting news on Twin Cities radio at the time.

On the first day of the all-news format on WWTC, listeners even heard commercials for competing radio stations WAYL and KEEY.

As the so-called "beautiful music" stations in the area at the time, WAYL (93.7 FM) and KEEY (1400 AM) figured they could lure listeners just finding out about the format change on WWTC by airing commercials on 1280 suggesting that disappointed listeners tune to their station. WWTC was willing to run the spots as long as the rival stations were willing to pay the regular ad rates.

Much to the chagrin of Lee Zanin, however, KEEY also ran newspaper ads headlined "Today WWTC radio trades violins for violence." With a large illustration depicting a gun pointed at the reader as the head of a tuxedo-dressed violinist, the ad went on to inform that beginning today, "WWTC radio replaces every bit of music with news".

Suggesting that "reports of Mideast tension, recession and crime" just doesn't soothe the nerves the way relaxing music does, the reader is assured "What you are about to give up on WWTC is what we play all the time on KEEY."

While the general manager of KEEY told a local media columnist he got "super reaction" to the ad, Lee Zanin told the same columnist reaction he heard concerning the ad was "100 percent negative".

The concept of all-news radio, while not for everybody, is certainly an asset to any radio market and has proven successful when a station is willing to shell out the money to keep the service interesting enough to keep people tuned in.

During the 1960s, stations WMIN (1400; later KEEY-AM) and KDAN (1370) made ill-fated attempts at news radio. WWTC was perhaps more determined to make it work. Unfortunately, the dullness of the NIS format and WWTC's lack of budget for a real local news operation made success painfully difficult.

It is primarily local news that has made other all-news operations successful. People tend to be much more interested in what is happening in their own back yard than what is happening on the other side of the world.

Minneapolis Star columnist Jim Klobuchar likened WWTC and NIS's version of all-news radio to "taking part in an all-day parade as a passenger in a revolving cement mixer"; the same stories, especially on a slow news day, were repeated over and over. The lack of earth-shaking events at every moment of the day makes for an awful lot of repetition, especially for a news station with the financial incapability to do any of its own research.

WWTC's leading competitor would be the mammoth WCCO, whose audience had always been too large for anyone to seriously tap in to. Bob Pirro, of the trade publication *Insider* compared the two by calling WCCO the "giant" and WWTC the "pygmy".

As time went on, WWTC didn't see anything dramatic ratings-wise; but the format did begin to catch on. Listeners who tuned in tended to stay tuned longer than most music radio listeners because some news reports did have the ability to hold interest, especially on days when things were really happening. Also to 'TC's advantage was the fact that news radio listeners are more alert to what they're listening to, therefore commercials have much more selling power. WWTC was able to raise advertising rates every few months as a result.

However, in November 1976, just as the format was catching on, Lee Zanin was informed that the NIS service would be folding in mid-1977. Affiliates had dropped it in some cities and it just didn't have the financial support to continue.

Zanin was not happy about this; the inexpensive NIS service provided WWTC with one-half to two-thirds of its programming by this time. The loss would be hard but Zanin was determined to keep WWTC an all-news station. When the announcement of the NIS discontinuation was made, Zanin wasted no time in preparing to fill the wide-open program gap. In fact, Zanin saw at least one advantage to the pending demise of NIS; the station would be able to commit itself to more local news coverage.

The first thing he did to make up for lost programming was subscribe to United Press International's wire service coverage in addition to that of Associated Press, which the station already had.

Next he went to the parent company of NIS and acquired the basic NBC Radio Network affiliation. With the NBC feed, WWTC was supplied with five and a half minutes of news at the top of each hour and about a dozen daily features presented by the likes of David Brinkley, Joe Garigiola, Edwin Newman and Bess Myerson, plus Notre Dame football and "Meet the Press"on Sundays. WWTC soon affiliated with the Mutual Broadcasting System in addition to NBC. The station also became home to play-by-play coverage of Minnesota Kicks soccer and Minnesota Gopher hockey, and later, Gopher football.

Lee Zanin in 1976. **(©1997 Star Tribune/Minneapolis-St. Paul)**

He went on to hire a dozen new employees including producer Dave Hellerman, sports director Doug McLeod, engineer Steve Brown, and reporters Greg Magnuson, John Farrell, Bob Bundgaard and Karen Youngquist-Riley. Most of the new hires, although experienced in news to some degree, were under the age of 30.

The offices and newsroom were remodeled to equip the enlarged staff and the station, according to Zanin, spent some $50,000 to equip the main studio to handle interviews with up to six people at once and add a third broadcasting booth.

"We had a pretty big boiler-room situation going on", recalls John Farrell. "We had a 24-hour newsroom. We had a person working overnights gathering news, we had a producer who set up what we called 'live-lines', which were essentially just telephone interviews. We had people coming in during morning drive, we would have reporters on-call who would handle breaking news like early-morning fires, things like that. We usually had [at least] two to three people behind the scenes in the newsroom at all times so it was a pretty busy place."

But staff and budget inequities continued to plague the station, so a "tip line" was established, enabling listeners to help the station out by calling in news tips. As a reward, tipsters received a $25 savings bond.

News radio WWTC did acquire something of a following in the Twin Cities, if only a small one. St. Paul Mayor George Lattimer was an avid WWTC news radio listener as were members of the Minnesota State Senate and other "movers and shakers" in the metro area.

On May 29, 1977, as the News and Information Service signed off into oblivion, WWTC continued its commitment to all news, all the time.

Talk Back Radio

Of the 68 radio stations that were still affiliated with NIS when the network folded, 42 planned to continue all-news programming. Lee Zanin met in Dallas with representatives of the other 41 stations to look into forming their own network of national news programming and features.

The network never did materialize but WWTC found another way to feature something unique — call-in talk shows.

Talk-back radio programs were popular in the turbulent sixties and in the Twin Cities, the talk-radio station of the time was WLOL-AM (1330). In the comparatively dull seventies (long before the Rush Limbaugh renaissance), call-in programs lost their appeal and there were few left on Twin Cities radio until the spring of 1977 when WWTC hired talk show veteran Bob Allard to do a call-in program from 10:00 p.m. to 1:00 a.m. six nights a week called "TC Talkline".

Allard began his broadcasting career as a pageboy for WCCO Radio. From there he worked at other radio stations throughout the upper Midwest and he was a news anchor on KMSP-TV, Channel 9 from 1961-64. He was a part of WLOL's news-talk format as a popular but often controversial host until 1973, before he finally resurfaced at WWTC.

Allard was soft-spoken but aggressive. He delighted in provoking people, not by yelling or pounding his fist, but by more subtle means. He knew how to get under people's skin. He would start his program each night read-

Bob Allard in 1978. **(©1997 Star Tribune/Minneapolis-St. Paul)**

ing random headlines from his favorite newspaper, the *Minneapolis Star*, and entice people to call by saying "if these phone lines don't fill up, I might have to read the entire newspaper — including the classified ads". He seemed to dislike children and would rudely hang up if any kids called in.

Describing himself as a pro-choice liberal, one of Allard's favorite topics of discussion was abortion. He would frequently bring the issue up, prodding listeners who disagreed with him to call. There were frequent on-air arguments over when life begins and he would demand that abortion opponents who called in spell out what penalties should be imposed on women who seek abortions and on the doctors who perform them. He openly admitted on the air that he was trying to antagonize the "anti-choice people".

In one typical spar on Allard's show, recounted in a *Minneapolis Tribune* profile, a caller insists "Life begins with the fetus and goes through the aged."

Replies Allard, "You never had the experience...of sitting down to eat an egg fresh off the farm, or maybe not so fresh, and you found out it had been fertilized. It was still a very edible egg.
"

"Correct..."

"Well, you were sitting down to fried eggs, not fried chicken."

"Biologists will tell you", the caller continues, "that if something is growing, it's living. And the fetus is growing from week to week."

"I see", Allard replies. "So if I go to the barber and get my hair cut I'm guilty of severing off some living entity from by body, those curly, golden locks. And they tend to grow back, don't they?"

"No. Haven't you heard about hair transplants?"

"That's not my problem, Madam. It may be yours. Thank you and good night."Allard moves on to another call.

With the success of "TC Talkline", WWTC began "TC Sportsline" with host Ed Cain on February 20, 1978. "Sportsline" aired Monday-Friday from 6:30 to 8:30 p.m.

Minneapolis Tribune sports writer Tom Briere described "Sportsline" as "argumentative, controversial, informative and progressive." The host described it as "50-50 informational and public participation".

Legendary North Stars coach Lou Nanne made frequent appearances on 'TC Sportsline.
(Courtesy Art Phillips)

Ed Cain had been a sportscaster on KSTP-TV, Channel 5 from the late sixties until 1971. From there he went on to several other ventures including calling play-by-play for the Mets games on WOR Radio in New York.

Cain had sports figures as in-studio guests. The premiere broadcast featured North Stars coach Lou Nanne. Everyone from Calvin Griffith and Rod Carew to Roy Smalley, Bud Grant, Cal Stoll and even O.J. Simpson were guests. (The only ones to turn down the show were the late Billy Martin and hockey's Derek Sanderson.) On one program, he interviewed an FBI agent about sports gambling. Listeners were invited to call in and talk to the sports figure in the hot seat or discuss and debate sports with the host.

Like Allard's show, Cain's "Sportsline" could stir up ire. His wrestling spoof angered promoter Verne Gagne and he was denounced by the Minnesota North Stars when he criticized the team's ticket price increases.

Two-way talk was catching on at WWTC. Soon Allard's producer, Dave Hellerman, was hosting his own call-in show, from 1 to 3 p.m. weekdays and Saturdays. Still in his twenties, Hellerman's wit had a youthful tone, sharp and intelligent and often irreverent.

WWTC resumed broadcasting on a 24-hour basis in early 1978, after broadcasting on a 20-hour schedule since going all-news. The wee-hours were filled with "The Larry King Show" from Mutual Network. Larry

King, who followed Allard, hosted a national call-in show with guests in the first two hours and open phones in its third hour. The program continued its run on Mutual until 1993 as King went on to bigger and better things.

WWTC remained all-news in the morning with "The Morning Report" from 4:30 to 10 a.m., a potpourri of network and local news, traffic reports, sports, weather reports and features and commentary by the likes of Jack Anderson, Cleveland Amory, John Erlichman and humorist Laird Brooks Schmidt.

WWTC was described on the air as "24 Hour People Radio" and the station even had the audacity of calling itself "The Twin Cities Most Important Radio Station".

Talk-back radio on WWTC provided the Twin Cities with unique radio programming in a market saturated with music stations. Even KSTP-AM (1500), the long-time talk station in the area was a top-40 rock station throughout the seventies. Call-in shows were also an advantage over strictly all-news programming for 'TC because they're less expensive to produce, they're far less redundant and they attract a large, intelligent audience.

The new format enabled 'TC to offer listeners something WCCO wouldn't: controversy. While the Good Neighbor to the Northwest served milk and cookies to every politician who passed through town and made a point of being heartwarming and homey, WWTC allowed room for conflict, debate and talk with actual substance.

Riot on Second Avenue

Many newsmaker guests appeared on WWTC talk shows but the most controversial by far was a 23-year-old American Nazi named Steven Martinson of the National Socialist White Peoples Party. His July 14, 1978 appearance on Bob Allard's program caused a near riot outside the WWTC studio.

Days before Martinson's appearance, Allard received calls during his program telling him he should not "allow" Martinson on the air. When one caller called Allard a "capitalist creep", Allard replied, "Well frankly, that's not how I run this show so please go to hell", and hung up.

About 20-25 protesters, many from the Progressive Labor Party, an avowed communist organization, picketed the WWTC building on Second Avenue on the Friday night of the scheduled appearance. A woman named Alida Bakuzis spoke over a megaphone while protesters carried yellow and red flags. Police were also in the vicinity.

When Martinson and his entourage arrived at about ten p.m., they were met with taunts and catcalls from the protesters, who blocked the entry. Soon fists were flying and a melee broke out. Police moved in and used teargas and nightsticks to control the crowd. Seven or eight were arrested. A large window at the Dolphin temporary service office, located at ground level of the building, was broken before Martinson finally got in, with a torn shirt and bruised arm.

"Our goal is to make America an all-white country, to insure the survival of the white race", Martinson told a bemused Allard on the air. "We plan on doing this, not through armed revolution, but through the ballot box."

All five WWTC talk lines were jammed with calls that night. As callers probed and debated him, Martinson insisted there was no Holocaust and, upon being asked about the Diary of Anne Frank, that there was no Anne Frank. He blamed Jews for abortion, homosexuality, race-mixing and even communism, among other things, and said that when his party takes over, all blacks would be shipped to Africa and that American Indians would be confined to the reservations.

Insisting that the wrong side won World War II, he speculated, "if Adolf Hitler would have won, I don't think we would have any black crime, we wouldn't have any pornography, we wouldn't have drugs...the after-effects of World War II are very obvious."

The bizarre but memberable broadcast was commented on for days by WWTC callers. The woman who spoke in the megaphone during the protest, meanwhile, was charged in Hennepin County District Court with rioting a week later.

A New Man In Charge

Effective September 1, 1977, longtime WWTC general manager Lee Zanin left to join the institutional bond firm of Dain, Kalman and Quail (now Dain Bosworth). He was succeeded by 28-year-old Doyle Rose, who, like Zannin before him, was promoted from the sales department. Rose was remembered as being more "jovial" than Zanin by at least one WWTC news staffer.

Shortly after Zanin left, news director Ken Trimble, known for his frequent tantrums in the newsroom, departed, much to the relief of staff. WWTC reporter Bob Berglund replaced Trimble.

Also in September, Minneapolis millionaire Robert Short agreed to buy WWTC from Buckley Broadcasting (formerly Buckley-Jaeger) for $600,000, pending approval by the Federal Communications Commission.

Buckley was in severe financial trouble and seemed desperate to get the property off its hands. They were selling it for a mere $100,000 more than they had paid for it thirteen years earlier. It was reported that the station had lost $157,892 the previous year, making it the least profitable station on the Buckley chain.

The FCC officially approved WWTC's ownership transfer to Short's company, Metropolitan Radio, in early 1978 and Short took over WWTC effective May 16, 1978. Short held 50.87 percent interest with his seven children holding 7.01 percent apiece. Son Brian would be overseer of the business. Under the license transfer agreement, Short was to continue the current format and maintain the news staff, which consisted of 17 by this time.

WWTC staff was jubilant over the sale. Being aware of Short's wealth, it was hoped that he would breathe new life into the station and that as a locally-based owner, he would care more about its day-to-day operation. Disappointment came all too quickly.

Robert Short was something of an infamous figure in the Twin Cities. Although he was born and had lived in Minneapolis his whole life, and claimed unadulterated loyalty to his community, he was the one who purchased the Minneapolis Lakers basketball team in 1957 only to move it to Los Angeles three years later.

He built his fortune on a trucking firm he had founded and later purchased other diversified businesses, mostly based in Minneapolis. His only experience in broadcast ownership, however was with an FM station in Los Angeles, which he owned for about five years in the 1960s. He owned a large part of of downtown Minneapolis, with a reputation as a shrewd but merciless businessman.

Something seemed a little curious about Robert Short taking over news radio WWTC; he was a Democrat Farmer-Labor candidate for U.S. Senate in 1978 and some questioned if he was going to use the station as a tool for his campaign.

"We were pretty high profile in a rather small circle of people who worked in the governmental apparatus and for that reason, Bob Short may have felt that this would be a useful commodity in his effort to buy the Senate race", speculates former reporter John Farrell.

A week before officially taking over WWTC, Short was running commercials for his candidacy on the station but quickly removed them when he realized that an opponent might demand equal time and that a conflict-of-interest situation could arise. But if an independent committee that happened to support Short ran spots, it would look a lot better.

Engineer Al Flom recalls, "He made up a fake support committee for himself and they [Short and son Brian] came to the station to record fake 'Committee Supporting Bob Short' commercials and putting them on the air. It was illegal as hell and I told Brian 'You're really stupid coming here to WWTC, the station you own, to record this stuff. You're just leaving a paper trail." The ads quietly disappeared from the air.

A Message From Above

As soon as Bob Short took over WWTC, he made his presence known to everyone who worked for the station.

Two letters by him were tacked to the station's bulletin boards. One said he would not accept any special coverage for his Senate candidacy (undoubtedly because the so-called "fairness doctrine" regulations of the time wouldn't allow it).

The second one reminded everyone of the station's "dreadful financial picture" and asked the news staff to "join the five members of the sales staff in a common sales effort". In other words, news reporters and gatherers were expected to drum up advertising on their own time, for ten percent sales commission.

The letter continued, "Some question has been raised that there might be a conflict of interest between reporting the news and selling commercial advertising on news programs. Common sense would indicate that any such suggestion is patently fallacious on its face."

Short concluded his message by saying "In the event any feel that they cannot lend a willing shoulder to this difficult wheel, I would respect their decision to withdraw from this common effort at an early date."

Newsroom employees were stunned. For news people to sell advertising, a question of ethics comes in; how fair and accurate would a story be reported if it could upset business interests? Short's closing comments also made many fear for their jobs, believing that if they didn't sell advertising, they could be fired.

Soon after the notice was posted, Short called a staff meeting, introducing himself to his employees, literally, by telling them that he was disappointed in the job they've been doing.

Reporter John Farrell remembers the meeting vividly. "We were called in, those of us who were actually in the office at the time, and he simply told us that we've been doing a 'dismal' job. He was a real cheerleader. He made it plain to us that he felt we were responsible for the station's failure. As employees we had dragged the station down and it is now necessary for us to redeem ourselves by hitting the bricks and selling once we got done with our various duties within the news department."

Farrell dared to speak out. "I don't think it's conceivable that people are going to take us as a serious news entity", he recalls telling Short. "On one hand we're supposed to report on them, on the other we're demanding that they buy advertising. There's an obvious conflict of interest there."

Short would not stand for such insubordination. He fired Farrell on the spot, making an "example" of him for anyone else who might raise objection.

Farrell gathered up his belongings and went home. But later that afternoon, he received a phone call from general manager Doyle Rose, asking him to come back to work. Rose had informed Short that they didn't have anyone else to do the morning news. He was back on the air the next day.

An anonymous 'TC employee told *Minneapolis Star* columnist John Carman that "before it was awful", but now Short is "treating us like totally dispensable items".

Short himself told Carman "We don't have any Cronkites over there", adding "I don't want anyone to work for me who thinks he's so high and mighty that he can't put his shoulder to the wheel".

Bob Short in 1978.
(©1997 Star Tribune/Minneapolis-St. Paul)

A few days after Short's policy was made public, Ernie Schultz, president of the Radio and Television News Directors Association, rebuked Short in a the trade publication *Broadcasting.* Schultz said the policy opens the door to an "impossible conflict of interest that will destroy the credibility of radio and television news".

Meanwhile, WWTC employees sought representation from the American Federation of Television and Radio Artists, AFL-CIO (AFTRA), not only because employees felt a lack of job security if they didn't sell ad time but also because some employees were reportedly earning less than three dollars an hour (minimum wage was $2.30).

WWTC was union in the days when the station was playing music. The announcers were with AFTRA and the engineers were with the International Brotherhood of Electrical Workers. But the IBEW had been voted out several years earlier and AFTRA left with the disc jockeys.

On July 31, 1978, the staff voted 13 to 5 to join the union. All 19 eligible employees voted, but AFTRA challenged the ballot cast by news director Bob Berglund, contending that he was management, thus disqualified. Negotiations began a few weeks later but a contract wasn't ratified for several months due to turmoil within the station. The station was falling apart at the seams.

Short resented the fact that the employees were organizing and successfully got the staff to vote the union out two years later, after a complete turnover in personnel. Ironically, Short's candidacy for Senate was advertised as "Labor Endorsed", perhaps for the simple reason that he was running as a Democrat.

No News Is Good News

As time went on, WWTC continued to sink deeper in the ratings and morale worsened every day. The Shorts were becoming impatient with the operating losses they were suffering and with the constant battles with the staff and management. They decided it was time to pull the plug on news-talk radio.

General manager Doyle Rose, who, unlike the Shorts, actually knew something about the radio business, pleaded with them to stick with the format at least until the new ratings data came out in January, and then evaluate the situation.

On November 29, 1978, Doyle Rose was dismissed. He was offered a sales position but he declined, taking a job instead as general manager of WLOL AM-FM. After bringing a top-40 format to the FM a few years later, he turned WLOL into the number one music station in the Twin Cities, and was eventually hired to head the parent company, Emmis Broadcasting. But he wasn't good enough to work for Bob Short.

"We were guaranteed", says John Farrell, "when he bought the radio station that we would have some time to rebuild it as a news entity. There was going to be this fresh fusion of cash.

"As I recall, the rational behind changing his mind on that score was that somehow or the other he'd been flummoxed into thinking that the station was on more solid financial ground than it was. On the basis of that he was going to go back on his word to us about keeping the format in place."

Doyle Rose was hired by WLOL to replace Wayne (Red) Williams — who coincidentally had become the new general manager at WWTC.

In his decade or so at WLOL, Williams had changed format numerous times on the AM and FM stations, never achieving much success with either. With that track record, he was hired to make WWTC successful.

By the end of 1978, news director Bob Berglund and "Sportsline" host Ed Cain left the station. Reporter Bob Bundgaard replaced Berglund who eventually joined Rose at WLOL. Veteran sports announcer Frank Butel took over "TC Sportsline". Butel, who was an old friend of Red Williams, had been a voice at radio 1280 in the '50s when the station was still WTCN.

The only thing that was really keeping WWTC afloat was Bob Allard's "Talkline". Allard was drawing some 11,500 listeners per night while the station's quarter-hour average throughout the rest of the day was only 4,300. However, those who listened to Allard and other WWTC programs tended to be older males, a demographic not considered to be particularly attractive to advertisers.

The ever-agitated Bob Short was losing money on a daily basis and was at his wit's end. He felt that he couldn't get any cooperation from his employees and he had little knowledge of how to run a radio station. He hired the Los Angeles-based consulting firm of Gold & West for advice.

The consultants told Short that the potential audience for news radio was too narrow and suggested a more low-risk format of "adult contemporary" music. "Adult contemporary" was the most widely used radio format in the country so with its proven success, the consultants assured, WWTC can't go wrong with it. Red Williams was given the task of implementing the new format.

The Splendid Blend

In an attempt to, in the words of the general manager, "broaden the base appeal from the listener's point of view", Williams began to dismantle the news-talk format in February 1979, shifting the station's focus to what was called "full-service adult-contemporary", consisting of pop hits along with information and other non-musical elements. Several other stations in the Twin Cities had a more or less identical format.

The first thing he did was pull the plug on Bob Allard's "TC Talkline" program, in spite of the fact that Allard had more than twice the audience of any other program on the station. Ironically it was Williams who fired Allard from WLOL-AM in 1973 when that station dropped its own news-talk format. WWTC also dropped news and features from the NBC Radio Network but retained the Mutual Network.

Spice with bright personalities tuned in on the pulse of the Twin Cities.

Start with the latest news: all of it, and all up-to-the-minute.

Add a dash of sports: colorful sports reports, plus generous portions of play-by-play

Start cooking with the greatest record hits of all time.

Fold in time and weather updates, plus helicopter traffic reports.

That's WWTC 1280 AM; radio the way it should sound.

1280

Blended Splendid

ALL HOME & AWAY KICKS GAMES BROADCAST ON RADIO - WWTC 1280 AM

TWIN CITIES MAY 1979 25

Being WWTC's most popular program, Allard's cancellation drew numerous complaints. Some even suspected the show was canceled because of his stand on abortion and his on-air antagonism of pro-life organizations such as Minnesota Citizens Concerned for Life, which station owner Bob Short supported. Both Allard and the station denied that was a factor.

After leaving WWTC, Allard took his program — and his 11,500 or so fans — to KSTP-AM (1500) and later to WAYL-AM (980). After being canceled on WAYL, the 55-year-old Allard took his own life on May 31, 1982.

WWTC began playing music on February 26, 1979. The new format was dubbed "The Splendid Blend", consisting of a "blend" of music and news. Short didn't want to keep the current air staff around but he realized that if there were massive firings, unemployment insurance claims would have to be paid out so he kept the staff, most of whom had little or no experience outside the realm of news, on board in hopes that they would just get frustrated and quit.

The displaced news reporters and call-in hosts who continued as disc jockeys included Tom Myhre, Greg Magnuson, John Farrell, Dave Hellerman, Doug McLeod and news director Bob Bundgaard, who was renamed program director.

Hired consultants suggested the new format but they didn't bother supplying the music so the station had to build its music collection from scratch. The records from the old music format had long since been sold or given away. Station staff was asked to bring in whatever records they had at home that could be considered "adult contemporary" music, although nobody was really sure what the term meant. The music was taped on cartridge along with a few current hits purchased at a nearby record store. That became the WWTC music library.

John Farrell remembers, "A lot of [the music] came from my record collection and the collection of [chief engineer] Al Flom...I brought in Lee Dorsey's 'Workin' In the Coal Mine', stuff by the Guess Who; I would also sneak in stuff that didn't belong there like 'Run Run Run' by the Third Rail or something by the Mojo Men or the Chocolate Watchband." (Everything, including those more obscure songs would remain with the "Golden Rock" format.)

Not caring much about their new duties as disc jockeys, some of the air staff mocked the role. Farrell was sometimes calling himself "Johnny Dollar", a name that was used by at least one DJ in every city in the fifties and sixties, or "your Uncle John on the radio". Doug McLeod, who was only doing weekend shifts, jokingly called himself "Robert W. Sunday" after the numerous imitators of Los Angeles disc jockey Robert W. Morgan.

Says Farrell, "It was pretty chaotic. I kind of enjoyed it in a way, once I knew that all-news was dead, I was free to sort of just goof around for a few months until I got a job somewhere else."

The "Splendid Blend" included some features from the old format including "TC Sportsline", Mutual's "Larry King Show" and Mutual's hourly news and features. WWTC continued to carry Minnesota Kicks soccer, the 1979 State hockey and basketball tournaments and Minnesota Gophers football.

Just as the new format got off the ground, "Sportsline" host Frank Beutel, who had been appointed by the Minnesota Kicks to call play-by-play for the games to be aired on WWTC, felt there would be a conflict of interest situation if he continued to host the sports talk show, so to the chagrin of station management, he quit "Sportsline" after only a few months of serving as host. He was replaced by veteran sports announcer and noted sports memorabilia collector Gene Harrington.

WWTC was still lacking any direction. AM radio was rapidly losing ground to FM in the late seventies, the format change alienated the news-talk listenership and morale around the station was so low the sales staff was reportedly unmotivated to sell advertising. The station's attempts at mixing talk and sports programming with music also worked to its determent, not to mention the constant use of the outdated word "splendid". The spring ratings showed WWTC to have just six-tenths of one percent of the overall listening audience, down two-thirds from the previous 1.7 percent.

Almost Shut Off the Air

In May 1979, Brian Short, son and business partner of station owner Bob Short gave the unmotivated sales staff an ultimatum. He told them he was not satisfied with the way things were going and that "If we don't show some improvement by the first of September, one of my recommendations to my father is that we just shut the thing off".

The sales staff was motivated. They began to aggressively sell ad time and by the summer of '79, the station was showing some small signs of improvement.

Brian Short didn't carry out his threat and the Shorts would have had a difficult time just shutting down the station, because of federal regulations. But while this was happening, two non-profit organizations began swarming over WWTC.

The operators of religious station KUXL (1570) negotiated with Short about moving that station to the stronger signal at 1280. Minnesota Public Radio (MPR) was also interested in relieving the Shorts of their burden.

MPR, which operated KSJN-FM (91.1) in the Twin Cities as well as five other non-commercial outlets throughout the state at the time was interested in acquiring a troubled AM station for its news and informational programming, leaving the FM station open for more classical music programming. WWTC was a definite candidate.

Brian Short told the *Minneapolis Tribune*, "Every time a ratings book came out [MPR] sent my dad a letter". Bob Short had talked with MPR officials but the terms weren't right for him. MPR eventually acquired WLOL's AM station, known at the time as "13 Rock" WRRD, turning it into KSJN-AM, later KNOW. Just over a decade later, MPR took over WLOL-FM (99.5) as well.

Positudes and Negatudes

As the months went by, "splendidly blended" WWTC became less defined. Initially the station's entire music library consisted of about 150 selections on tape cartridge, played in high rotation. Because of the lack of direction and management, some disc jockeys were playing their own records from home on the air. The format was originally intended to be "adult contemporary", i.e. soft pop hits, but the "splendid blend" ended up including hard rock, disco and fifties music as well.

Station overseer Brian Short acknowledged that the only way to make the faltering AM station successful is to do something nobody else is doing and do it well. Adult-contemporary wasn't cutting it.

In the summer of 1979, the Shorts dropped the consulting firm of Gold & West and hired Florida-based Burns Media. While program director Bob Bundgaard jumped from what was perceived as an already sunken ship, Burns Media sent Minnesota native Scott Carpenter, who had been programming a station in Florida, to try and reign things in. The Shorts liked him and immediately snatched him up as permanent program director.

"My primary job was to put some cohesiveness to it", Carpenter remembers. "They were having trouble with [WWTC], I said 'Let me go in and take a look at the thing'. And I said 'Oh my God!'"

The place was a mess. The programming was completely unstructured, nobody seemed to know or care what they were doing and the morale amongst the employees was downright ugly.

"Everyone's looking at me like, 'here's a new son of a bitch'", he recalls. "Everybody was just so full of passioned hate, they didn't know where they were going...I realized when I came in and accepted the job that I had a huge morale problem."

Carpenter, with his decidedly off-center personality and his unorthodox management theories, was viewed as a bit unreal to the staff. He had a philosophy major and was a firm believer in the power of positive thinking. He gave motivational speeches that some staffers weren't sure what to make of.

"I don't know what he was using but he had pupils the size of dinner plates", comments John Farrell. He recalls a staff meeting where Carpenter was "standing up at this chalk board with all these people lined up at any side of the table and like a football coach he's drawing these swooshing arrows from one point to the other talking about 'positudes' and 'negatudes'. . .everybody was looking at each other like 'Christ, I hope this guy doesn't have a gun'."

Whatever differences the staff might have had with Carpenter or the Shorts or anyone else, Carpenter decided to do something about the morale problem. He told everyone to meet him at the Court Bar, the popular watering hole around the block from WWTC, bought them food and drink, and made them a deal.

"You guys, you're pissed, you don't like it here, you don't know what your future is, you're news people, I've done news so I feel your frustration. A lot has been mismanaged so far", he told his bemused staff.

"I tell you what: you give me as much time as you can, give me your best performance possible and in return, I'll help you find work and I will allow you to use WWTC to make tapes for your next job. I will give you as much time as I can. The only terms will be if you get real negative and start causing problems, then the deal's off, on an individual basis." Everyone agreed and shook hands on the deal.

"It became cohesive, it became positive because these guys did follow through. I was really pleased", says Carpenter. "They were frustrated but they were willing to put their frustrations aside. We shook hands on the deal and we got into certain positives, positive feelings and it worked."

Meanwhile, Carpenter dropped the incredibly out-of-date "Splendid Blend" tag line and WWTC began calling itself the home of "The Twin Cities Greatest Hits". The oldies rotation increased and he began bringing in new blood to WWTC.

The old news-turned-DJ staff began to drift away. John Farrell did his last show on WWTC on September 15, 1979, gleefully telling his listeners, "That's gonna do it for your Uncle John on a permanent basis, I won't ever see you again. Dave Hellerman is next and you'll forget all about me in no time." He punctuated his farewell with a sardonic laugh.

"I left because I didn't want to work for Bob Short, I didn't want to work for Red Williams, I didn't want to work amid that chaos", Farrell says. "WWTC was really kind of the bastard stepchild from the beginning."

Farrell worked for music station KCLD-FM in St. Cloud briefly before returning to his first love, news, at KSTP-AM. Hellerman soon departed as well, joining former boss Doyle Rose at WLOL, as had former news director Bob Berglund. Hellerman also eventually turned up at KSTP, doing an afternoon call-in show, similar to the one he had once done at WWTC. Doug McLeod and Deanna Burges also left within the month. By October, the old air staff was completely gone.

"TC Sportsline" and "The Larry King Show" were dropped. WWTC kept Mutual news but dropped most of the other network features except Mutual sports updates, which came over the wire daily at 1:35 p.m. and a feature called "Eye on Hollywood".

A whole new air staff was phased in and WWTC became "The Golden Rock of Minneapolis-St. Paul". The goose had finally laid the golden egg — or so it would seem. While everyone was optimistic about the station's success, Carpenter would leave in less than a year. He realized all too soon that under the management of the Shorts, the station would never really be successful.

WWTC had finally seemed to rise from the dead. But the freewheeling party did in fact hit rough terrain.

WEDNESDAY, JANUARY 20

THE GOLDEN ROCK **WWTC** 1280 AM *AND* **CITY PAGES**

INVITE YOU TO THE RETURN OF THE "SURFIN' BIRD"

THE TRASHMEN

SPECIAL GUEST

B.J Crocker and the "T.C." Sound of Music Machine

GOLDEN OLDIES DRINK SPECIAL

8:00 to 9:00
B.J.Crocker Special
Free Tap Beer

8:00 to 10:30
WWTC Golden Rock Special
50¢ Bar Drinks • 50¢ Domestic Beer

12:00 Midnight
The Trashman Special
FREE FREE FREE FREE FREE
The Famous Trashcan Toddie
Five great boozes prepared & served just the way you like it—in the trashcan!!
PLUS
Many Surprises All Night!!

Doors Open at 7:00 $3.00 Cover Music at 9:00

4. THE GOLDEN ROCK ROLLS DOWN

Second to 'CCO?

At the time Bob Short took over WWTC in 1978, he told *Minneapolis Tribune* columnist Neal Gendler "I see no reason you shouldn't be number two in the marketplace. You're never going to beat WCCO but you can be second...with a little more imaginative programming".

For a while in the early eighties, WWTC seemed to be well on its way of doing just that. Even with little presence in regards to promotion (billboards, TV spots, ads), WWTC could be heard in a lot of places throughout the Twin Cities. At traffic lights during the summer when car windows are rolled down, it wasn't uncommon at all to hear the sound of golden oldies blaring from AM car radios. TC played in the intercoms in several convenience stores around the area and even the bicycle-powered radio at the Science Museum of Minnesota was set at 1280, rewarding the person who peddled the stationary bicycle hard enough to turn the radio on.

At night when the WWTC transmitter changed direction and the signal traveled and bounced around throughout the atmosphere, the station received phone calls and letters from Canada and beyond.

A student at the University of Minnesota named Roger Bull and his two roommates had parties with the Golden Rock blasting away, having name-that-tune and name-that-artist contests as part of the action. WWTC so inspired Roger, he decided to pursue his education at Brown Institute, the famed broadcast school in Minneapolis. His first radio job was at WAVN, a tiny daytime-only AM station in Stillwater. He later moved to Duluth's country music powerhouse, WDSM.

While working in Duluth, he received a call from WWTC program director Mike Ryan, who was acquainted with Bull, offering him a job with the Golden Rock. Ecstatic, Bull headed back to the Twin Cities and submitted a resume and demo tape to the station. He was hired to do the 7 to midnight shift, six days a week. His first task, however, was helping move office equipment from the old Second Avenue studio to the new studio on the 12th floor of the Wesley Temple Building at 123 East Grant Street in downtown Minneapolis. Everybody on the staff had a hand in that.

The entire staff was jubilant about the new home of WWTC. "I remember how excited we were about the Wesley because it was much larger", says Nancy Rosen. "We had a teeny-weeny studio on Second Avenue and the Wesley Temple was this big, huge thing. [The studio] was this big, huge room with windows that looked over Grant Street and you could see the skyline of the city. It was beautiful because on Second, we had no windows that looked outside."

The move was thought to be yet another step up for the rapidly accelerating radio station; but by the time the station was finally settled in at the new facilities, something in the atmosphere seemed to change.

The Party's Over

Almost everyone who worked for WWTC when it moved to its new home in the Spring of '81 notes in retrospect that something seemed different in the new facilities; something almost spiritual was left behind with the old building on Second Avenue.

"When we moved to Wesley it got to be more serious", recalls Mike Ryan. "Down on Second, we could get by with Ugly Del calling from phone booths and Brad Piras having coffee slurping contests in the morning...but then we got to Wesley and it seemed to be, like maybe we should become

more serious about this instead of trying to continue having fun. It took a different atmosphere at that point."

Observes Arne Fogel, "When that station moved to the Wesley Temple Building, it lost a huge something or other, I can't put my finger on exactly what it is but it ceased to be a big-time radio station in some way."

At least one thing stayed the same. "All the shit equipment we had they took along and I was doing a morning show with everything set on top an old wooden desk", remembers Del Roberts, "and they said they had no idea if they'd ever get new equipment or build a new studio."

He also recalls, "the roof leaked and when it rained the water poured on the electrical equipment. I'm surprised someone wasn't electrocuted."

Adds Nancy Rosen, "The equipment was very shitty. This little teeny board that was about this big with rotary pots. It was easy to work with, I wouldn't say easier than what they are now but there wasn't much to work with so what could be hard about it?"

The Shorts eventually invested in new equipment and building maintenance several months later.

WWTC's success was waning. After reaching a peak Arbitron rating of 3.9 in the Spring of 1980, the rating slipped to 3.5 in the fall and the trends were looking shakey for the next ratings sweeps. It was the beginning of a downward spiral.

The one area where TC was looking good was in the morning with the "Ugly" Del Roberts show. Del was holding up well in spite of the giant morning audiences of WCCO-AM and KS95-FM and if nothing else, Del was one of the most popular and recognizable DJs on WWTC.

Unfortunately, he was becoming a bit of a thorn in the side of a few people at the station. He has a reputation for being strong-willed and as a result, he irked a few people.

Ugly Del Bids Farewell

Del Roberts' morning show on WWTC got good ratings but on the week of May 18, 1981, in the middle of a ratings sweeps, the Incredibly Ugly Del disappeared, abruptly replaced with veteran announcer Barry Siewert, who was known as Barry McKinna on KDWB from 1968-1973 and was the voice of WTCN-TV "Metromedia 11" throughout the seventies.

The official management explanation for Del's departure was that he was just a temporary replacement for the recently resigned Brad Piras and that Siewert was the permanent morning DJ. Management left open the possibility that Del may return for future fill-in duties.

But there was a lot more behind Del's sudden disappearance. John Carman, *Minneapolis Star* broadcast critic and Golden Rock fan let the worms out on the can in his May 29, 1981 column. Carman revealed that Del had clashed with station management over the selection of music he played on his show. Del played what he considered appropriate for early-morning listening; primarily early sixties tunes along the lines of Dion, Gene Pitney and Rick Nelson. Management felt that a variety of selections from different eras should be played. There had also been complaints that Del had been playing the same certain songs every morning.

Eventually station bosses had music director Arne Fogel put together a playlist every day and ordered Del to play the selections listed in sequence.

According to Carman in his article, "one morning that meant he had to play a gruesome old song called 'Dinner With Dracula' at 7:45 a.m. He also recalled such awkward music shifts as a segue from a Jan and Dean surfing record to a John Lennon ballad 'Imagine'".

Recalls Arne Fogel on the matter, "It wasn't my idea to make out that playlist every day for Del. It was a waste of my time and I knew that Del would hate me for it, which was exactly what happened."

The *Star* column didn't exactly endear station management or others to Del, who tipped off Carman, either. Fogel says he was "incensed" when the article came out. "I was fuming when I read it."

Del also told Carman, who revealed in the article, that his pay at WWTC "was about $4.50 an hour, less than he earned at KDWB-AM (630) in 1965."

The bosses at WWTC were not happy about that revelation but that was the one part of the article Fogel says he had no objection to. The low pay was a sore spot with everyone within the station. Operations manager Dick Driscoll, who dealt with the owners regularly, frequently tried, with little or no avail to get pay raises for everyone.

"The disc jockeys were grossly under paid on the air", Driscoll admits. "Brian Short never understood, and to this day I don't think there is a Short who understands people.

"When I was trying to get raises for the disc jockeys, I talked to the Labor Department and found out the average immigrant Hmong was making two dollars an hour more than our disc jockeys were. Nothing against the Hmong people of course and I'm glad they were successful...but [Brian] Short never did understand people because he never had to work for a living."

In the end, Del was not tapped for any more fill-in duties at the station. "All I know is with what I did I got the highest ratings of anybody on the staff but nobody would listen to me so what the hell", he says in retrospect.

The family was breaking up. With the departure of Del and of Brad Piras several weeks earlier, Nancy Rosen moved to Winter Park, Colorado, where some friends were trying to start up their own radio station. The operation never did materialize so she ended up taking a job at a top FM station in New Mexico.

With Rosen's departure, Mike "Records" Ryan, who was named program director, took over the seven-midnight evening shift, while continuing to work as a daytime personality at a small AM station in Stillwater. He quit the Stillwater job after receiving an ultimatum by WWTC management.

Roberts, meanwhile, who had dabbled in several diverse careers including law enforcement, became the police chief of an Indian reservation in South Dakota; but his friends at WWTC hadn't heard the last of him yet.

The Sky Is Falling!

In June 1981, the Spring Arbitron report came out. WWTC had slipped from a 3.5 to a 3.2 rating. It was a small decrease, certainly something that should be dealt with, but the station owners hit the panic button. They were convinced that if some kind of drastic action isn't taken, the station would be bankrupt in no time.

According to Dick Driscoll, the Shorts thought that when 'TC jumped from a 0.6 to a 3.9 rating within a year, they had a golden goose on their hands. It's only a matter of time before we take 'CCO, they thought.

"So when it went down, even a tenth of a point, it was like Chicken Little — the sky is falling! The sky is falling!"

Brad Piras, the former program director who had left the station a few months earlier, had warned the owners of the potential ratings problems the station was facing and had proposed a few minor changes such as cutting the commercial load, cutting some of the morning talk features and upgrading the quality and selection of new music. His ideas were scoffed at and the frustrated Piras soon quit. Now the station didn't have him to blame or to remedy the problem.

Arne Fogel, perhaps unfairly, was blamed somewhat for the ratings decline because he was the music director. Fogel, with his rather mild personality, was quite often general manager Charlie Loufek's whipping boy. While Loufek received tongue-lashings from the owners over the ratings decline, Loufek took it out on Fogel.

Instead of putting more money into the operation for promotion and other things that might attract new listeners, the Shorts opted instead to hire yet another west coast-based consulting firm to figure out what was wrong and try to come up with a remedy.

The Shorts had hired consulting firms two years earlier with mixed results. It was, ultimately, consultants that created the Golden Rock format. Since that particular firm had parted company, the format had drifted somewhat away from the original concept. A few minor adjustments, perhaps, were needed.

Common sense might tell someone taking a good look at this radio station that the steady ratings decline of the past few periods might be due to the fact that FM was becoming more serious competition for AM radio in general or that the decline could just be a natural fluctuation or it might have something to do with the fact that WWTC dumped its most popular air personality during the recent ratings sweeps. The hired consultants from out of town had their own ideas.

WWTC's ever-expanding music library had grown to well over two-thousand songs on tape cartridge, thanks in part to dedicated listeners and staff who lent the station their own records. If it was released as a 45 RPM single, chances are WWTC played it. The consultants saw that as problem number one. The first thing they advised was to pull everything that didn't make *Billboard* magazine's top ten out of the play rotation.

Using Joel Whitburn's *Billboard Book of Top-40 Hits* as the official guideline, the station was told to weed out anything that didn't make the top ten — regardless if it made the top ten locally but not nationally.

It went very much against the grain of program director Mike Ryan to heed the consultants. "We literally cut the library at least in half", he recalls. "If it was an eleven, you couldn't play it. That got frustrating."

The consultants, in all their wisdom, recommended that Art Phillips, a 19-year-old weekend DJ, be fired because of his youth and lack of experience. After his dismissal from WWTC, he was hired by WCCO-FM (102.9), eventually getting promoted to WCCO-TV (Channel 4) where he remains as a producer, camera operator and engineer.

Operations manager Driscoll battled with the consultants and the owners who hired them, telling them "You can't program this thing like a rock station does where you're figuring on an audience turnover every 20 minutes and the music turns over every 30 days. They're not making any new 'golden rock' tunes."

Neither the consultants nor the owners listened, and the modified Golden Rock was implemented. Arne Fogel was relieved of his duties as music director but remained on the staff as a weekend personality, producer and copy writer.

By the Fall of 1981, WWTC sounded noticeably different; it seemed to have become an AM station pretending to be an FM station.

The personality aspect was cut along with the music selection. Tuning in to 1280 no longer meant tuning in to a wild party where almost anything could happen. There was no more dedication line, no more off-duty disc jockeys paying surprise on-air visits, no more prank phone calls, and the joking around and wisecracks were cut down to a minimum.

Thanks to the consultants, the TC airstaff became civilized and professional. Three or four songs in a row were played with no talk between them. Listeners want to hear music, not talk, the consultants insisted, pointing to research from radio stations in various cities.

After each selection of songs, the announcer came on with what one veteran radio personality calls "before-that syndrome"; i.e. "1280 WWTC, you just heard the Beatles doing 'Paperback Writer'. Before that you heard Dion and the Bellmonts with 'Runaround Sue'. Before that you heard..."

Also, with two-thirds of the library gone, the DJs had to tell numerous disappointed listeners that they no longer played that favorite song they just requested.

Observes Fogel, "It appeared to me that first of all, [the consultants] had absolutely no concern with the market. They didn't care that they were in a specific market that is different from other markets. The philosophy was, this is what worked in Kankeekee or whatever, it's gonna work here. Second of all it seemed to me that they were taking the notion of how to set up an oldies station where there hadn't been one rather than looking to sustain the ratings at this particular one that already exists."

Meanwhile, Mike Ryan became so frustrated having to work with the consultants that he resigned as program director, staying on as an announcer. Dick Driscoll took over as acting program director as the airstaff turned over some more.

The Changing cast of Characters

By late 1981, the voices heard on WWTC had shifted around. Barry Siewert, the morning drive host who replaced "Ugly" Del Roberts early in the year, moved on to spin country music on WDGY (1130). Overnight man Steve "Boogie" Bowman took over the morning show.

While Dick Driscoll did double duty as operations manager and program dircctor, he was scarcely heard on the air at all. Mike Ryan moved from the evening to midday slot while B.J. Crocker continued doing the afternoon shift. Newcomer Roger Bull took over in the evening.

Bull had enrolled in Brown Institute aspiring to work for the Golden Rock. He got his wish rather early in his broadcasting career. In the first hour of his show each night, Rogcr played a full hour of fifties and sixties "make-out" music. "That was not my idea", he is quick to point out.

"Dick Driscoll thought that one thing everybody just loved about the oldies was the love songs so he made sure there was an hour of love songs programmed...so here I was, following B.J. Crocker who was like, Mr. Rock 'n' Roll Deejay, screaming and yelling, and his screaming and yelling would end and we'd go into Roberta Flack, but I had to live with it".

Roger Bull
(Courtesy David Carr)

Following Roger was Nancy Gallos, who did the overnight shift in place of Bowman. Nancy was the daughter of WCCO-TV personality John Gallos. She had a "girl-next-door" quality to her, a little more mainstream than Steve "Boogie" Bowman had been in that time slot.

The station added Don Thompson and Shaun Waggoner to the weekend lineup. Thompson, the current resident engineer at WWTC, was a bit older than the rest of the airstaff. His career began at WCAR Radio in Pontiac, Michigan, where he announced the Detroit Tigers games. His on-air style was loud and brassy, resembling that of local TV pitchman Mel Jass. On top-of-the-hour station breaks, he'd give the time by bellowing out "Ladies and gentlemen it's nine o'clock!"

Waggoner did the overnight shift on weekends. Unlike Thompson, Shaun, who almost never used his last name on the air, sounded mellow and spaced-out, resembling former KQRS "stereo underground" personality Alan Stone from the late sixties and seventies.

Clap For the Wolfman

On Saturday afternoons, WWTC ran the syndicated "Wolfman Jack Show", featuring the legendary gravel-voiced DJ playing golden oldies "that make you wanna HOWL!" along with interviews with singers and musicians in a three-hour national program.

The howling persona of the late Robert Smith loved the Twin Cities and would occasionally do a live local show on WWTC. Dick Driscoll remembers, "Wolfie used to come to town every six months or so and spend all day Saturday on the air with us. He was usually promoting Wolfman stage packages of some kind or the other, in which he brought in a number of name acts to the auditorium. So we'd let him promote that and he'd talk to our listeners when they'd call in.

"Wolfie would say things like 'Hello, Wolfman here! Are yer peaches sweet?' Well nobody knew what that meant exactly but it sounds slightly obscene. It used to utterly confound our owners. Short called up and said 'What does he mean by that? He's talking to some teenage girl on the air. What's going on?'

"I said 'It means nothing, Bob. She was shopping at Lund's and he was just checking out the produce department.' I'm not sure that [Short] bought it but I guess he decided not to push it any further."

The station also ran promos featuring the Wolfman's distinct voice. "Hi, this is Wolfman Jack", he said in one, "remindin' you that yer listenin' to the Golden Rock; 1280, double-yuh double-yuh tee cee!"

In another spot, the Wolfman said of the Golden Rock, "now this is REAL RADIO"! Curiously, after that spot had ran on the station, WWTC received a call from WCCO Radio asking if WWTC intended on using "real radio" as its official slogan. WCCO was told the station had no plans for it.

Lo and behold, WCCO was soon promoting itself as "Real Radio", even copyrighting the slogan. WCCO is quick to point out that the late Senator Hubert Humphrey, who worked closely with 'CCO throughout his political career, had called the station "real radio" but those who were a part of WWTC in the early eighties like to gloat over the fact that mammoth WCCO had to go through wimpy little WWTC before they could register the phrase.

The Sky Is STILL Falling!

Broadcasting can be a cruel business. It revolves entirely around ratings based on research provided by companies like Arbitron, Birch and Hooper. People are hired, fired, promoted and demoted based on what a few hundred out of a community of a couple million say they listen to. Advertising dollars, i.e. profit, depends on ratings.

The owners of WWTC spent several thousand dollars hiring a consulting firm to aid the station's slumping ratings. The consultants advised some rather drastic changes in the station's oldies format and the expensive advice was heeded.

When the ratings came out for the Fall '81 sweeps period, WWTC went from a 3.2 to a 2.5 overall rating, the biggest drop the station had seen yet. The consultants were promptly fired.

Sensing opportunity, former program director Brad Piras, who had warned the station that this would happen, came out of the woodwork. He called general manager Charlie Loufek in early 1982 and told him he was back in town. He said he would like to come back to WWTC and Loufek was interested.

But there were strings attached. Piras wanted to return as program director and morning personality just as he had been before and, as part of the deal, he wanted to finally try out the programming proposal he had offered to the station shortly before leaving. Loufek responded by naming B.J. Crocker as program director. Piras decided to pursue other business ventures.

With Crocker as p.d., more changes were made within the station as 'TC tried desperately to turn its situation around. He brought most of the 2,000-plus selections of the music library back into regular airplay, adding even more to it with the help of a record collector friend named Bob Broz and station intern Stuart Held. B.J. encouraged the disc jockeys to be creative again and brought back a few of the gimmicks that had made the station popular in the past, such as the theme weekends, where songs of a particular subject or some other common factor were spotlighted.

Mike "Records" Ryan, frustrated and feeling he had done about all he could do with WWTC, departed to take a job as program director at WAYL-AM (980). With Ryan's departure, Roger Bull moved from evenings to the late morning-early afternoon shift. Nancy Gallos took over the evening shift, and began doing a nightly all-request show from 6 to 11 p.m.

Bull had been doing a request show on Saturday nights, which proved to be successful enough for the slumping radio station that Crocker decided to make it a nightly feature.

Less than two months after being appointed as host of the evening request show, Nancy Gallos was bumped back to late nights and the request show was taken over by John Messenger, a stronger air presence who had previously spun records for KDWB (630) and KSTP (1500).

Along with Messenger, WWTC hired the soft-spoken Carol Lasota on a part time basis. The unassuming 22-year-old who only did weekends quickly acquired her own "groupies", much to her amusement, including a couple of young men who claimed to be the "Carol Lasota Fan Club". She had recently graduated from Brown Institute, the Minneapolis broadcasting school, and considered working in radio to be "just a hobby".

While morale at the station seemed to be improving, things still weren't improving on the business end. While general manager Charlie Loufek and operations manager Dick Driscoll denied persistent rumors of a possible format change at 1280, there were numerous discussions between management, the owners and various advisers as to how to get this format back on the right track.

It was decided that the station should try to attract a somewhat younger audience. WWTC's Golden Rock was most popular with the over-35 set. It also did well among 18-34 year-olds, an age group advertisers tend to prefer, so 'TC decided to cater more to that younger group.

More psychedelic-era '60s and early '70s music was added to the library. Vividly colorful melodies such as "Time Has Come Today" by the Chambers Brothers, "Foxy Lady" by Jimi Hendrix and "Hot Smoke and Sassafras" by Bubble Puppy were becoming a regular part of the Golden Rock.

'TC also promised to play something "new in '82" once an hour, upgrading the selection of new music. Whereas before, an occasional pop single by Kenny Rogers or Carly Simon was the extent of TC's "new music", they were now playing a more progressive selection of current hits to attract younger ears. (Ironically, this was one of the key elements of Brad Piras' proposal which had been so soundly rejected by management.)

Harder pop and rock songs such as "Paperlate" by Genesis, "The Other Woman" by Ray Parker, Jr., "Gloria" by Laura Brannigan, "Feet Don't Fail Me Now" by Todd Rundgren and Utopia, "We Got the Beat" by the Go-Gos and "65 Love Affair" by Paul Davis were among the many "new in '82" hits TC was playing along with the oldies, none of which would have been heard in 'TC's new music rotation had they come out a year or two earlier.

The *Twin Cities Reader*, the hip "alternative" weekly newspaper, named WWTC as "Best Pop Radio" in their 1982 "Best of" issue, "for consistently playing today's classics today — be it Prince, Musical Youth, Toni Basil, or even Bananarama...as well as their always super selection of past hits."

Of course, this is exactly what was making many of the dedicated Golden Rock fans tune out; there were plenty of contemporary hit stations on the dial but only one oldies station.

Promotions Gone Haywire

Even the thing WWTC did best, on-air promotions and rock shows, turned into disasters in at least two instances.

In the earlier days of the Golden Rock, DJ Mary Hatcher was the station's unofficial promotions director. She was the brain behind a lot of the contests and on-air bits that gave the station its charm and personality. Del Roberts, Brad Piras and Scott Carpenter also came up with many of the fun elements of the format.

By early 1982, however, they had all departed as did so many other staffers and the station didn't appoint anybody else as promotions director. Some staffers were coming up with their own promotional ideas for the station but for the most part, the station wasn't doing nearly as much along those lines, primarily because the owners often refused to finance any major contests or advertising campaigns, outside a few billboards and bumper stickers.

One of the more disastrous low-budget contests the station put on was "Arne Fogel's Create Your Own WWTC Jingle Contest". It was general manager Charlie Loufek's concept and Fogel wanted nothing to do with it from the start. The contest centered around him because he was the quasi-celebrity of the staff, having recently released a local hit record, and because he was, by profession, a jingle writer and singer.

In this contest, participants were to compose an original jingle for WWTC and send the tape to Arne, who was to judge the best one. The winner won "The Fogel Award".

Ads for the contest were rather deceptive. WWTC had no intention whatsoever of actually using any of the jingles submitted on the station. Fogel was very weary of being in the forefront of this fiasco. "I felt very self-conscious...I sort of knew what was going to happen; we're going to get a tremendous public response, I'm the public face behind this thing and I'm going to get no support from the station and everyone's going to end up pissed off at me; and that's exactly what happened." On top of everything else, Fogel had to listen to and judge the contest entries on his own time.

Some of the tapes he received were laughable. One that he recalls was of someone simply saying into a tape recorder "W-W-T-C/Look-at-me". Others weren't as bad.

When Arne chose his winning jingle, he had to shop around on his own time and money to find a cheesy little trophy for his winner. Being in charge of this entire contest which he wanted nothing to do with in the first place, he also had the responsibility of telling his winner the "Fogel Award" was all he was getting.

"It was a dumb Charlie [Loufek] promotion that paid no attention to hurting people's feelings or creating bad will", comments a still somewhat bitter Fogel. "That created bad will and bad blood."

A much bigger disaster that created even more bad will involved WWTC's sponsorship of a Wolfman Jack rock 'n' roll oldies show at the St. Paul Civic Center.

Wolfman Jack, the well-known howling disc jockey was a staple at WWTC. His syndicated oldies show was heard Saturday afternoons on 'TC and he did commercials and even live shows on the station while visiting Minneapolis.

'TC naturally sponsored the Twin Cities stop on the Wolfman's national concert tour. He was emceeing an oldies show that included the Regents, the Shirelles, the Bellmonts and the Grass Roots on the bill. The concert drew a full house at the St. Paul Civic Center on that Friday night in May 1982 but there was turmoil brewing backstage.

There had been a dispute for some time between WWTC and J.R. Productions, the company putting on the show, and promoter Hank Troje, Jr. A year earlier, WWTC had aired some $10,000 worth of advertising for Troje's last show which hadn't been paid for. WWTC had taken Troje and the production company to court at the time where Troje claimed he didn't order the advertising that ran on the station, so the station agreed to a $3,000 settlement. The $3,000 still hadn't been paid up.

On top of that, Troje and J.R. Productions were having Wolfman Jack do spots for the WWTC-sponsored show on competing radio stations. Station owner Brian Short was especially incensed by this.

Short ordered the station to immediately cancel the Saturday afternoon Wolfman Jack radio show and obtained a court injunction to tie up the box office funds at the concert to pay off the money owed to WWTC.

On the night of the show, as the Civic Center filled up, agents representing Short's organization walked in armed with a court order that legally tied up the funds at the auditorium. The acts performing could not be paid nor could any money be distributed until all claims against the production company and the promoters were settled. As a result, the show was delayed for more than an hour.

While Short's agents demanded the money and the producers argued and negotiated back stage, Wolfman Jack tried to keep the increasingly antsy audience warmed up. Short didn't want to stop the show; that would mean admission would have to be refunded and the producers couldn't have raised the money to pay the station.

The show finally did go on but not quite as planned. The Shirelles, not knowing if they would be paid, did not perform. A member of the audience recalls an on-stage scuffle while the Bellmonts were performing. According to the eyewitness, a man in a business suit came out and attempted to yank one of the musicians off stage, grabbing him by the shoulder and saying "C'mon, get off the stage right now!" The musician continued playing. When the businessman tried grabbing him again, another man in a business suit came out grabbed the first one and the two men in business suits broke into a shoving match before going behind the curtain.

In the aftermath, there was bad blood between those who paid to see the show and WWTC, who came out looking like the bad guys in the ordeal, as well as bad blood between the station and the producers of the show. It was also the end of Wolfman Jack on WWTC.

Hot Fun In the Summertime?

The summer of 1982 kicked off with a WWTC "World's Greatest Class Reunion" (by this time there were two "class reunion" parties per year, one in spring and one in fall, plus a St. Valentine's Day party, a Halloween party and a Christmas party) and a fun-for-the-whole-family "pizza pigout" at

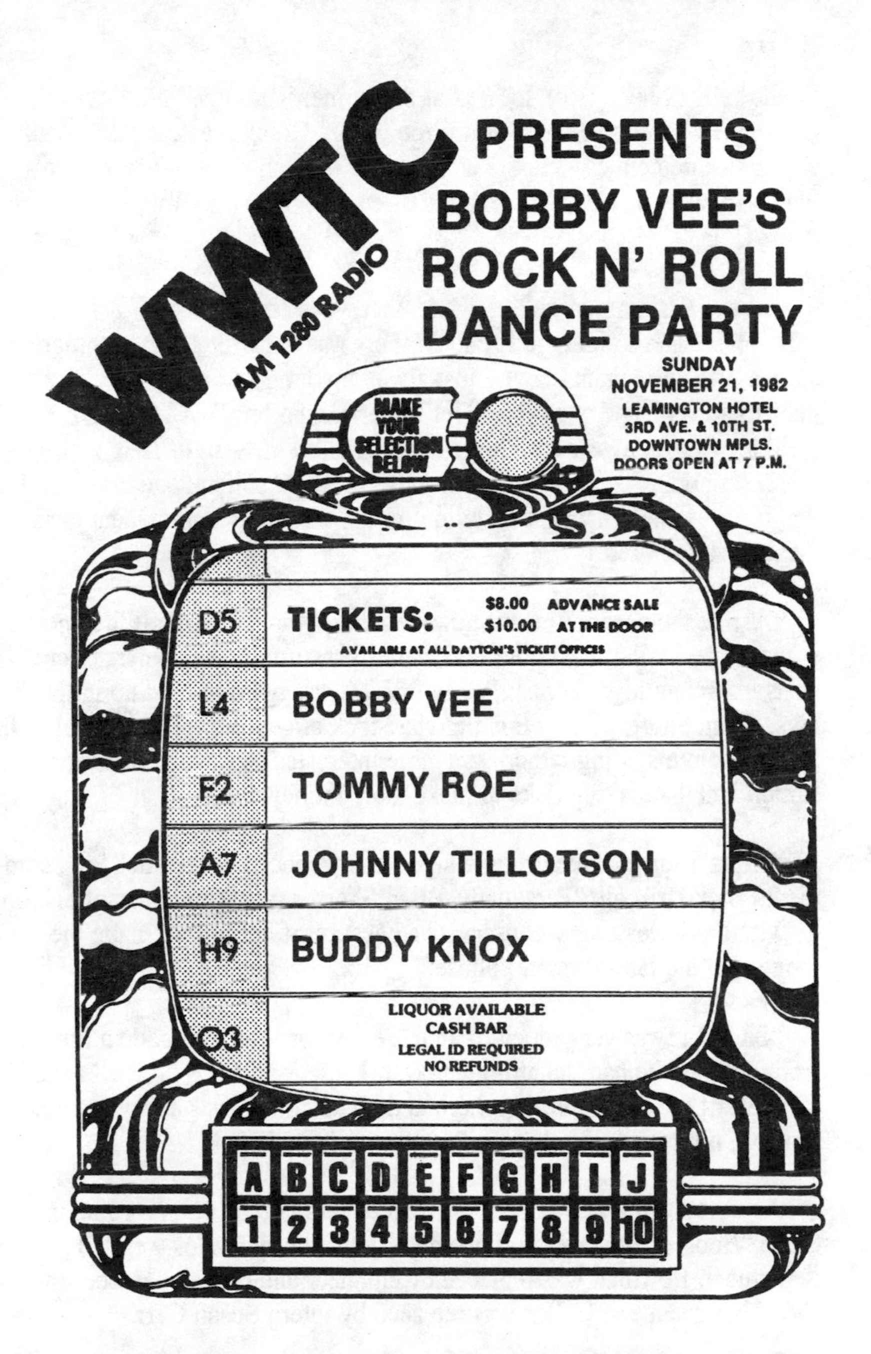

The WWTC dance parties continued to be the best show in town-but the listening audience was melting away. **(Courtesy Art Phillips)**

Waldo's Pizza Joynt (sic), located at the former west-side Porky's drive-in near Lake Calhoun in south Minneapolis. B.J. Crocker emceed the event which featured clowns, classic cars, hot air balloon rides, Willy Wanka's magic show (sans Gene Wilder) and of course, plenty of Waldo's pizza and Pepsi-Cola.

For the first time in quite a while, there was optimism at WWTC. John Messenger's nightly all-request show was becoming very popular, the station was once again playing just about anything and everything it could get its hands on and even Ugly Del Roberts, who left WWTC on not-so-good terms over a year earlier, paid a special on-air visit to 1280 in July, 1982, sounding just as wild and crazy as ever. (In spite phone calls and letters from listeners asking the station for his return on a permanent basis, Del was not rehired.)

Despite the new optimism, however, there was still turmoil behind the scenes. The station was under continuous pressure by the owners to cut costs but retain high ratings. While WWTC struggled to hold up in the ratings, Brian Short, who by this time had took almost complete control of the station from his ailing father, was more interested in saving what money the station had than taking risks to make more money.

"Brian Short had a hard time signing any check over ten dollars", comments Dick Driscoll. "Personally, Brian Short is a charming person, is fun to be with but is very, very conservative because of not having made the money [of the family estate] himself.

"Short Sr. was very much of an entrepreneur type. He made a fortune from nothing, he had the attitude that 'if I lose it all I can do it all again'. The rest of the Short family inherited their money. It was always 'protect the money', not make more but protect what they had."

As a result of Brian Short's insistence that costs be cut, morning man Steve "Boogie" Bowman and overnight jock Nancy Gallos were let go of in September. Bowman was replaced by engineer and occasional weekend jock Don Thompson and Gallos was replaced by intern Susan Carr.

Bowman eventually moved to KEEY ("K102") FM where he did a late night country music request show. Gallos turned up on WCCO-FM, staying with the station when it became WLTE.

On July 23, 1986, Steve Bowman died of lung cancer at the age of 41. Rachel Bowman told the *Minneapolis Star and Tribune* of her late father, "He liked working the nighttime hours. A lot of times he'd get called by the same people every night and they would talk...sometimes I didn't like the music but I thought he was real good."

Steve Bowman

With the departure of Bowman and Gallos, Mike "Records" Ryan returned to the station doing weekend shows. During his six-month absence from 'TC, he had been program director at "easy-listening" WAYL-AM (980). A child of rock 'n' roll, the ever-frustrated Ryan returned to 1280, where he was greeted by numerous listeners who called in to welcome him back on his first afternoon on the air.

Roger Bull moved back to evenings, taking over the all-request show and John Messenger went to middays. The station also hired a new weekend jock named Alan Freed, although he was not related to the legendary New York DJ credited for coining the phrase "rock 'n' roll".

As Bull took over the request show, the number of requests coming in per night declined sharply. The request show was canceled but Bull remained on the evening shift.

Bob Short Dies

Early in 1982, WWTC owner Robert Short was diagnosed with terminal cancer. Mr. Short had a reputation for being tough-as-nails and often a controversial figure, but certainly a good, strong leader. The prospect of him not being around to guide the destiny of the station worried many at WWTC, especially since it seemed inevitable that son Brian would be running it when the senior Short was gone. Many questioned Brian's ability to run the station.

As Bob Short's health rapidly deteriorated, many decisions involving WWTC and the other businesses the Short family owned were left hanging. Short, for example, had the opportunity to buy KTCR-FM (now KTCZ) for two million dollars and another FM station in Montana. The Shorts made no final decision in either event so both deals fell through.

Dick Driscoll speculates, "If they had purchased what is now KTCZ for the two million dollars [The Golden Rock] would probably still be on the air today."

Bob Short passed away on November 20, 1982 at the age of 65. Mike Ryan, the disc jockey on the air that night, checked with the Short family before making any announcement on the air.

Mr. Short himself decided that he wanted no special memorial or tribute to him on the air but rather a simple announcement. So Ryan, known so well for his dumb jokes and dry wit, took a very serious tone of voice as he announced:

"The staff and management of WWTC Radio regrets to announce the death of Bob Short. Bob Short owned several businesses in the Twin Cities including radio station WWTC."

At the memorial service, held a few days later at the Leamington Hotel, Ryan was the only member of the WWTC staff who attended.

From there, Brian Short took primary control of WWTC.

A New Oldies Station In Town

As the year turned into 1983, a new station popped up on the AM dial which threatened head-to-head competition for WWTC.

KQRS, which for years had been simulcasting its programming on its AM and FM stations decided to do separate programming on 1440 AM. The format they chose was rock oldies.

KQRS-AM became KGLD, or "1440 K-Gold". The new station promised to play "all the hits from four decades of music!" K-Gold was run primarily by KQRS programmers Doug Sorrenson and Vicki Hodgson. Hodgson was in charge of promotions and research as well as being assistant program director.

1440 K-Gold, as it turned out posed no real threat to WWTC; it was hallariously awful. There were no live air personalities except for one in the morning. Otherwise the station was run completely on automation, whereupon a giant machine played all the tapes of music and commercials. Between every other song, a recorded voice came on announcing "1440 K-Gold plays all the hits!" Sometimes the machine malfunctioned and the tapes overlapped.

Dick Driscoll recalls, "We listened to it for about a week and thought this is a joke. Nobody in their right mind would listen to it for any length of time especially if we ran contests, had phone things, talked to our listeners; [we were] people who were really there, you could touch us, you could see us...live personalities will always run circles around automation."

When the first ratings book that included K-Gold came out, the station was dead last with just three-tenths of one percent of the listening audience. A year after the format went on, 1440 AM was simulcasting with KQRS once again.

Solid Gold 1280

While that other station was calling itself "K-Gold", WWTC was billing itself as "Solid Gold 1280 — the Original Solid Gold Music Station". 'TC was determined to out-gold the other guys.

Call letter jingles were added to the format. The station purchased a cut-rate jingle package for a mere $100. It featured a chorus singing the call-letters and dial position and sometimes a short verse such as "Driving along/singing a song/with a friend/W-W-T-C/twelve-eighty" and "Twelve-eighty W-W-T-C/bubbling over with/FUN! The "Solid Gold" theme was also prevalent in the jingles.

(Courtesy Del Roberts)

While the station was upgrading its sound in early '83, the airstaff was shuffled around once again. John Messenger left for KEEY ("K102") FM which had just changed format from easy-listening to country music. Roger Bull, who had been with the station for only a year and a half, was switched to Messenger's midday time slot. This was his fourth rescheduling. Mike Ryan got the evening slot and Alan Freed became the permanent overnight jock. Shaun Waggoner departed for KQRS shortly after.

WWTC attempted to be a vehicle for the increasingly trendy Twin Cities local music scene. At 7:20 p.m. weeknights, Mike Ryan did a segment called "Where The Action Is", where he would announce what local night-clubs the oldies cover bands such as Hitz, the White Sidewalls and the Rockin' Hollywoods were playing.

The station also launched an hour-long Sunday night feature known as the "WWTC Music Forum" with Arne Fogel as host. Fogel moderated a serious discussion on the local music scene with a four-person panel of musicians, club owners and bookers, critics and other figures in the industry. The primary focus was on bands performing new music instead of oldies bands.

"Music Forum" was the creation of Sue McLean of the sales staff. Mc Lean was very much into the "hip" music scene and was often a little disappointed in the not-so-hip Arne Fogel as host, but Fogel tended to just stand back and let the panelists lead the discussion.

Crackpots and Nut Cases

From the beginning of the Golden Rock, WWTC had not only its fans but even its fanatics as well, people who called every disc jockey on the station every day, made a point of attending every remote broadcast the station sponsored and even hung out at the studio.

The live remote broadcasts would always attract a crowd to a particular store or event and were very successful. The disc jockeys had fun doing the events but they often felt rather self-conscious being the center of attention.

As Roger Bull remembers, When I was out broadcasting live, people would come out just to say 'hi'. Every remote that I did the same 15 to 20 people would show up and sometimes they'd stay there for the whole two or three hours that I'm broadcasting. It'd drive somebody nuts sometimes doing that, you're just there trying to do your job and someone's just standing in front of you staring at you."

The "groupies" were usually nice, harmless types who perhaps had too much time on their hands but there was always the occasional kook who would come out. When B.J. Crocker was broadcasting outside a tire store in Hopkins, a skinny, hyper young man with curly hair calling himself "Randall Crandall" and claiming to work for 1440 K-Gold came by, chanting "WWTC sucks! K-Gold rules!"

A bemused Crocker played along with him. The young man then asked why 'TC won't give him a job. "I keep calling you guys but you won't hire me."

"We don't currently have any openings", B.J. told him.

"Well fire Alan Freed! He ain't worth shit!"

B.J. remained calm but was clearly irritated, replying "I don't think you're in a position to make that judgment."

The most obsessive Golden Rock fans were also known to come right to the studio, sometimes creating a tense situation.

One afternoon, a rather plain-looking young woman in her late teens or early twenties came to the station with a Spiral notebook. She was such a Golden Rock fan, she managed to log every song and every commercial the station played, 24 hours a day, for three days straight. It was a rather impressive accomplishment and she wanted to show her efforts to the WWTC staff, in hopes they could use her services for something around the station.

As the friendliest station in town, WWTC staffers were willing to take a look at her work and were genuinely impressed with all the effort she put in to it. B.J. Crocker and Dick Driscoll gave her verbal pats on the back for her effort and her interest in the station.

However, the station had no use, and there certainly wasn't room on the payroll for someone with the incredible talent for writing down what the station already did. But the young woman was reluctant to take no for an answer.

The next day, she returned, notebook in hand. She spent most of the day sitting in the lobby, writing stuff down in her notebook. The station brass tolerated her, although B.J. kindly suggested to her that maybe she'd like to just go home and relax and he assured her he'd call her if the station can use her services.

The morning after that, the same young woman was back once again, with the urgent desire to be needed.

Finally, a station official told her as politely as possible that she will have to leave the premises immediately. The woman stared back, hurt and in total disbelief. She then became very angry and began raising her voice and making a scene. The young woman soon found herself being escorted by security guards down the elevator and out the door.

Politics As Usual

In the cruel business of broadcasting, the amount of talent or intelligence or any other asset one possesses matters very little; the method used to determine how popular a particular personality is, is curious at best, considering that the entire business and people's careers revolve around the almighty ratings surveys.

"One of the very brutal drawbacks of radio is you live and die based on what appears to me anyway, to be a very arbitrary and capricious way of determining if you have any listeners", comments Mike Ryan. "Send out a bunch of pieces of paper to people, tell them to write down what they're listening to and if they say they listen to you, then you're good and if they say they don't, then you're bad, you're out of a job and I think that's very brutal."

Ryan, who had been in radio since the late sixties, knew this all too well.

He'd seen it happen to a lot of friends and it'd happened to him more than once. In the spring of 1983, it happened to him again.

Former Golden Rock personality Nancy Rosen returned to WWTC in April, 1983. Rosen had left the station two years earlier to take a job as program director at an FM rock station in Colorado and later moved on to the top-rated FM station in Albuquerque, New Mexico. Upon returning to the Twin Cities, she reunited with her old friends at the Golden Rock.

WWTC was more than happy to have her back; but to make room for Ms. Rosen, someone would have to go. She was given the six-to-midnight shift. "Records" Ryan was shown the door.

"Mike Ryan was certainly part of the family", comments Dick Driscoll, "but sort of the black sheep of the family. When it came time for budget cuts of if someone like Nancy came along, Mike was often the first to go." Ryan did return to 'TC briefly in August, 1983 and again in September, 1984 to do fill-in shifts, but was not rehired on a permanent basis.

In the two years of Rosen's absence, WWTC had grown and changed considerably. She noticed things to be much more structured, not quite the same place she knew before. She nevertheless fit it quite well and was appointed production director upon her return to the station.

Meanwhile, in its never-ending quest for an identity, the station experimented with new weekend talent in hopes of finding someone who could bring back some of the magic and excitement that once possessed the Golden Rock. Two vastly different personalities, Scott Stevens and Gary Rawn, joined the staff in the Spring of '83.

Stevens, who usually worked Saturday nights, was wilder than any jock the station had seen in a long time. He was loud, boisterous and maniacal on the air (although quite soft-spoken when away from a microphone). A veteran of the infamous Twin Cities hard rock station WYOO-FM ("U100") in the seventies, he was known for yelling "GO FOR IT!" and "WHOOP!" during songs.

In addition to the regular rotation of oldies, Stevens, a noted record collector, occasionally brought in his own, usually obscure tunes. He did a regular feature called "Weird and Unusual" where he would play a bizarre novelty song or two.

Scott Stevens seemed to renew enthusiasm for WWTC. He got more phone action than old standbys B.J. Crocker and Roger Bull when he was on the air and he developed a following with his listeners in a very short period of time. When he was scheduled to fill in for an ill or vacationing DJ during the week, he would get two to three times the calls from listeners that the regular DJ would get. But politics being what they are in that business, the rising star at WWTC was fired less than five months after being hired. Stevens turned up a few months later on album rock station "Stereo 101" (KDWB-FM).

Gary Rawn, the other part-timer hired in the Spring of '83, was a little more laid back than Stevens but he came off as a bit corny. Usually working Sunday afternoons, he frequently boasted on the air of being from California and often dropped names of celebrities he'd supposedly worked with.

He always ended his show by saying "God bless, love you all...and remember, in the beginning there was rock and then man invented the wheel."

When B.J. Crocker took his annual two-week vacation in July 1983, Rawn got the coveted afternoon shift, which had been originally promised to Stevens. The phone action died down considerably in that time period. His hours were nevertheless expanded.

Oldies In Stereo

By the fall of 1983, WWTC was in even more disarray. The ratings had been steadily decreasing for some time, going from 2.3 in the fall of '82 to 1.4 in the spring of '83. WWTC hadn't seen an increase of any kind since 1980.

Adding insult to injury, 'TC's worst nightmare came true when an FM station decided to go oldies. KJJO-FM (104.1) switched from country music to "20 Years of Rock 'n' Roll In Stereo" in September 1983.

"The New K-JO 104" as it was called was a little different from the Golden Rock, geared primarily toward 'TC's younger audience and KQRS's older audience. The music K-JO played was primarily rock hits and album cuts from 1964 to the present. Along with Paul Revere and the Raiders and the Dave Clark Five, K-JO played such album rock artists as Led Zeppelin, Aerosmith and the J. Geils Band.

K-JO's morning drive DJ was none other than Mike "Records" Ryan. The disc jockey who just wasn't good enough for the little AM station at 1280 was welcomed at the new FM oldies station. They even picked up the syndicated comedy series "Chickenman", once a staple of the Golden Rock.

KJJO's general manager was Mike Waggoner, father of former 'TC jock Shaun Waggoner. Dick Driscoll recalls once inadvertently telling Waggoner that anybody who did the oldies format on FM, even if it wasn't done well, could "blow us out of the water". Waggoner took Driscoll's advice and the WWTC operations manager's foot was firmly in his mouth.

Meanwhile at the Golden Rock, program director B.J. Crocker, under ever-increasing pressure from management to get the station back on an upward trend, tried his hardest to deal with the competition on the FM dial.

After doing the afternoon shift for almost four years straight, B.J. teamed up with Don Thompson in the morning to compete not only with Mike Ryan on KJJO, but with other morning teams such as Knapp & "Doughnuts" on KSTP "KS95" FM and WWTC alumni John Hines and Bob Berglund on WLOL-FM (99.5). The ever-boastful Gary Rawn, who had been doing weekends, was assigned to afternoon drive.

Nancy Rosen, having returned to 'TC with much fanfare, took a temporary leave of absence after five months and was replaced by Debbie Thomas. With the schedule shifting, more weekend personalities were hired including Larry Nelson and Mark Johnson.

THE TWIN CITIES BEST MUSIC
Survey of WWTC Listeners, 1-30-84 to 2-2-84

1. Just You And I - Buddy Holly
2. Unchained Melody - The Righteous Brothers
3. Oh, Pretty Woman - Roy Orbison
4. Mack The Knife - Bobby Darin
5. American Pie (Parts 1 & 2) - Don Mac Lean
6. Little Darling - The Diamonds
7. The Sounds Of Silence - Simon & Garfunkel
8. 96 Tears - ? and the Mysterians
9. In The Still Of The Night - The Five Satins
10. Can't Help Falling In Love - Elvis Presley
11. Old Time Rock And Roll - Bob Seger & the Silver Bullet Band
12. Rave On - Buddy Holly & the Crickets
13. You Send Me - Sam Cooke
14. Lightnin' Strikes - Lou Christie
15. I Only Have Eyes For You - The Flamingos
16. Surfer Girl - The Beach Boys
17. Billie Jean - Michael Jackson
18. Brown Sugar - The Rolling Stones
19. Money -Barrett Strong
20. Celebration - Kool & the Gang

Musical Chairs

When the fall '83 ratings came out, WWTC had fallen another tenth of a point. It was a smaller drop, but a drop nevertheless. B.J. Crocker, frustrated and defeated, decided to step aside as program director. He also wanted to spend more time with his family. The childless and single Roger Bull enthusiastically took over as program director, the sixth to do so since the inception of the Golden Rock format.

When Bull took over the position in March, 1984, the airstaff was shifted around once again. Crocker went back to afternoons where he was more comfortable and weekender Larry Nelson was teamed with Don Thompson in the morning. Gary Rawn went back to doing weekends briefly before taking a job at KJJO-FM. Alan Freed and Debbie Thomas were let go of, as Bull returned to the evening shift to free himself up for business meetings during the day and Nancy Rosen returned to the air once again, taking the 10 a.m. to 2 p.m. shift.

Bull hired several new part-timers, including Darrell Renstrom and Steve Garran, along with Randy Randall and Ray Walby, former personalities of KTCR-FM (97.1), which had recently become "The Cities 97".

Renstrom says that internal atmosphere at WWTC when he got there was nothing at all resemblant to the legendary "party atmosphere" that defined it in the past. "It was pretty quiet around there", he recalls. "It wasn't a real upbeat vive at all...I can't say it was mellow, just quiet."

The evening shifts were rearranged under Bull. His shift now went from six to ten instead of six to midnight. Newcommer Marianne Berreth did a four-hour shift from 10 p.m. to 2 a.m. and Mark Johnson spun oldies from 2 to 6 a.m.

Roger Bull was generally an optimistic, easy-going type and he took the task of program director with confidence. He became humbled, however, as he experienced the heat from middle and upper management that made B.J. Crocker finally get out of the kitchen — only now it was hotter than ever.

"I was in meetings with the general manager and the owner where the owner brought in this financial consultant who said 'shut the thing down, turn the power off. You are losing money'", recalls Bull. "Brian Short was saying 'I love this radio station, I want to make it work'. So we would come out of these meetings just whipped from side to side with the owners saying 'I want it to work' and the financial consultant saying 'I don't want it to work' and then Charlie [Loufek] and myself getting into meetings where [we figured] we have to make it work or else we're going off the air."

There was now serious discussion of taking a drastic step and changing format at WWTC, dropping the Golden Rock all together.

Ugly Del Returns

As things were looking dark and pessimistic at WWTC, a beam of light broke through the clouds in the form of a disc jockey who called himself the "Incredibly Ugly" Del Roberts. After a three-year absence, Ugly Del returned to the Golden Rock in an attempt to breathe new life into the format he initially helped create, doing weekend shifts and fill-ins.

Hearing Ugly Del on WWTC at this particular point was not unlike entering a time warp; not only was he a voice of the past but he refused to conform to the more structured, more sedate sound the station had taken in recent years. It sounded as if for four hours on a weekend, it was 1980 again at the Golden Rock. The 1980 'TC sounded vastly different from the 1984 'TC.

Ugly Del talked once again of cruising Lake Street and hanging out at Porky's. He opened up the "dedication lines", inviting listeners to call on the air and dedicate the next song. Dedication line was a feature long forgotten at 'TC. As he went into network news at the top of the hour, he'd announce "You're listening to WWTC, Minneapolis-St. Paul, it's four o'clock in the Twin Cities, for those of you in St. Paul, big hand on the twelve, little hand on the four."

On one memberable Saturday afternoon, he played the song "Heatwave" by Martha Reeves & the Vandellas and immediately played it again when somebody called in and "dared" him to. He played it a third time when somebody else "dared" him.

"That makes it a threeplay!" he declared. "If I play it four times I get into trouble!"

One evening while filling in for Roger Bull, he invited the listening audience to surround the Wesley Temple building with their cars and flash their headlights in time to "Wipe Out" by the Surfaris and some other songs with a heavy beat, creating a "poor man's light show". Dozens of cars surrounded the building that night as Del viewed the spectacle through his 12th floor studio window. It so happened that DFL Senatorial candidate Joan Growe had her campaign headquarters in the same building. Her campaign staff was on duty.

Recalls Del, "They suddenly saw dozens of cars flashing their lights into their windows. They thought the Republicans were up to something and called the police, which only increased the light show."

Ugly Del, in spite of his efforts, unfortunately came a little too late. Behind closed doors, the fate of the Golden Rock was decided by management.

The Golden Rock Rolls To a Stop

In October, 1984, program director Roger Bull went to each of the disc jockeys and told them that effective November 12, WWTC would be switching to an "urban contemporary" music format. The general reaction was of shock and surprise, but just about everyone who worked there could see it coming for some time.

The "urban contemporary" format WWTC was about to switch to consisted primarily of funky, electro-techno dance hits. The music the station would soon be playing would be the stuff heard in the trendiest dance clubs in town — Prince, Sheila E., Shalamar, Midnight Star and the like.

Del Roberts took the news of the format change the hardest. "What you want to do here is fad radio", he protested to Bull. "Doing a disco format now would be like doing an all-surf music format twenty years ago!"

"Ugly Del loved that oldies format more than anyone else", recalls Bull. "He loved it more than anyone that ever worked there including Steve Bowman or Brad Piras or anyone...and for me to take his friend and flush his friend down the toilet; it was a sad day for all of us but for him even more so."

As the staff was told of the format change, they were advised not to tell anyone until the station officially announced it to the public, but it was a poorly-kept secret. Almost as soon as the decisions were made, complaints, some irate, began trickling in to the station. The calls and letters increased daily.

On October 30, WWTC held its annual Halloween party at the Leamington Hotel. Despite the ratings surveys showing that the station's audience was dwindling down to almost nothing, the party sold out again, with the entire Great Hall of States ballroom at the Leamington packed with a standing room only crowd. Although the party had been scheduled some time before decisions to change format at WWTC were made, it came to be the unofficial farewell of the Golden Rock.

By November 7, WWTC announced to the media that effective November 12, the Golden Rock would be history and the new uptempo beat of urban contemporary music would hit the airwaves at AM 1280. In transition, the station began playing its entire music library chronologically, starting from the fifties and going to the eighties with the intention of staying there.

The final week was a hectic one. While the station played its collection of more than 3,700 songs year-by-year, the airstaff and office people were bombarded by negative phone calls and letters from Golden Rock fans. On the Friday before the format change, a 17-year-old 'TC listener who was well known amoung the staff as one of the station's "groupies" spent the entire day picketing the Wesley Temple Building, home of WWTC, all by himself, much to the dismay and chagrin of those around the station who wanted the format change to go as smooth as possible. He was greeted by Ugly Del and B.J. Crocker and snubbed by Roger Bull and Dick Driscoll.

As the Golden Rock sang its swan song at WWTC, Bull, Nancy Rosen, Larry Nelson, Darrell Renstrom, Mark Johnson, Carol Lasota, Randy Randall and oddly enough, Del Roberts stayed on with the new format.

B.J. Crocker departed for middle-of-the-road station KMFY (980; formerly WAYL-AM), Don Thompson and Ray Walby went to KRSI (950), weekender Steve Garran went to KJJO-FM (104.1), Marianne Berreth went to WAYL-FM (93.7) and Arne Fogel, whose role had been gradually reduced at WWTC for some time, continued as a free-lancer for advertising agencies and as a studio musician,, and began a weekly program on KLBB (1400) spotlighting the music of Bing Crosby. He also was a regular contributor of programs for Minnesota Public Radio.

General manager Charlie Loufek and operations manager Dick Driscoll retained their posts at the station. While Driscoll hadn't done any airshifts for WWTC in some time, his deep, resounding voice was used in station i.d.s for the new format.

On Monday, November 12, 1984 at six a.m., Larry Nelson introduced listeners to "Metro Music 1280".

THE TWIN CITIES WORST MUSIC
Survey of WWTC Listeners 2-4-84 to 2-14-84

1. Tiptoe Through The Tulips - Tiny Tim
2. They're Coming To Take me Away, Haa-Haa - Nepolian XIV
3. (You're) Having My Baby - Paul Anka
4. Yummy Yummy Yummy - The Ohio Express
5. I Write The Songs - Barry Manilow
6. Transfusion - Nervous Norvus
7. Muskrat Love - The Captain and Tennille
8. Does Your Chewing Gum Lose Its Flavor (On The Bedpost Over Night) - Lonnie Donegan & His Skiffle Group
9. Lovin' You - Minnie Riperton
10. Elvira - The Oak Ridge Boys
11. Beep Beep - The Playmates
12. Bette Davis Eyes - Kim Carnes
13. Surfin' Bird - The Trashmen
14. My Ding-a-Ling - Chuck Berry
15. The Candy Man - Sammy Davis, Jr.
16. Purple People Eater - Sheb Wooley

17. Bobby's Girl - Marcie Blane
18. Hello Muddah, Hello Faddah (A Letter From Camp)
 -Alan Sherman
19. Short People - Randy Newman
20. Tie Me Kangeroo Down, Sport - Rolf Harris
21. Please Mr. Postman - The Carpenters
22. Teen Angel - Mark Dinning
23. The Name Game - Shirley Ellis
24. Seasons In The Sun - Terry Jacks
25. The Streak - Ray Stevens
26. Torn Between Two Lovers - Mary MacGregor
27. The Monster Mash - Bobby (Boris) Pickett
28. A Boy Named Sue - Johnny Cash
29. Paper Roses - Marie Osmond
30. Witch Doctor - David Seville & the Chipmunks
31. Sweet Pea - Tommy Roe
32. Feelings - Morris Albert
33. Norman - Sue Thompson
34. King Tut - Steve Martin
35. Coming Up - Paul McCartney
36. Laurie (Strange Things Happen) - Dickey Lee
37. Brand New Key - Melanie
38. Puppy Love - Donny Osmond
39. Midnight At The Oasis - Maria Muldaur
40. Wooly Bully - Sam the Sham & the Pharaohs

Note: Because of the overwhelming response, this survey had to be expanded to forty songs.

WWTC AM 1280

ON AIR PERSONALITIES

LARRY NELSON

MORNINGS 6:00 AM - 10:00 AM

Larry Nelson with a morning drive assignment will wake you up with music in his style, voice and charisma that sets him apart from the crowd. Larry Nelson is certainly an asset to the airwaves of the Twin Cities. Wake up with Larry Nelson mornings on WWTC.

NANCY ROSEN

MIDDAY 10:00 AM - 2:00 PM

As a dedicated broadcasting professional with eight years of experience, Nancy Rosen is one of the most exciting and stimulating radio personalities anywhere. Nancy Rosen is sensitive, understanding and loves to meet and be involved with her listening audience.

ROGER BULL

AFTERNOON DRIVE 2:00 PM - 6:00 PM

Metro Music Programmer Roger Bull brings the best to your drive home in the afternoons. Roger Bull knows his audience and enjoys public appearances where he can meet his listener friends. A trained musician and singer with an avid love of sports and music rounds out his character. Roger's bits of information, personality and charm would certainly never be dull with Roger Bull as your radio companion.

BENJIE McHIE

EVENING 6:00 PM - 10:00 PM

Self pride, artistry and discipline are words that Benjie McHie lives by because he loves life. Benjie McHie, a charming and entertaining person with an evening assignment. He is perhaps one of the most talented persons in radio today. Enjoy your evenings with Benjie McHie.

DARRELL RENSTROM

LATE NIGHTS 10:00 PM - 2:00 AM

If you're an all-nighter, looking for the right mix of entertainment and good music for your weekday wee hours, tune in AM 1280's Darrell Renstrom late nights. WWTC's ·FRESH MIX, featuring the hottest music cuts put together by the top DJ's from Twin Cities area clubs starts out the show, followed by a blend of the most captivating selections of contemporary soul, R&B, and jazz mixed with fusion, reggae and the best music from local artists the metro area has to offer.

MARK JOHNSON

OVERNIGHT 2:00 AM - 6:00 AM

Whether you're up late after a night on the town, working the third shift, or just suffering from insomnia — share your early mornings with Mark Johnson. Along with the great music heard throughout the day on WWTC, Mark Johnson plays the latest tracks by artists from all over the world. He also keeps you informed on what's happening in the world of music. So tune in at 2:00 a.m. to Mark Johnson — you may never want to go to sleep again.

5. ALWAYS NEW, ALWAYS FRESH

Metro Radio

Bright and early on the morning of November 12, 1984, disc jockey Larry Nelson inaugurated the new sound of 1280 AM.

"This is Metro Music WWTC", he announced as he segued into a funky record by Rockwell. Gone was his morning partner Don Thompson and the 3,000-plus songs of the beloved Golden Rock. No more oldies for this station; WWTC had become, like, totally now.

"This format is a recognition that there truly is an urban scene worth identifying with", program director Roger Bull told David Carr of the *Twin Cities Reader*.

Added music director Mark Johnson, "We want to reach out and play the new stuff, but always remembering that we're in the Twin Cities and our station is a part of the community. We look upon it as really continuing people's musical education."

"Metro Music" was based on the primarily black-oriented urban contemporary radio format that had proven highly successful in several major cities, but with a difference; Metro Music also included a strong emphasis on new music by locally-based bands, the kind of music that was played in the trendiest dance clubs in town such as First Avenue, William's, Norma Jean's and Payne Reliever. Along with playing the music, WWTC did live broadcasts from the clubs.

What listeners now heard at 1280 on the dial was a loud, pulsating beat with plenty of synthesizers and drum machines. The artists were more important to the format than specific songs. The listener heard album cuts and alternate versions of hits rather than just the hits.

Among the nationally known artists heard on the new WWTC were Chaka Khan, Madonna, Billy Ocean, Midnight Star, Tina Turner, Shalamar and of course, Minnesota's own Prince.

His movie "Purple Rain", filmed in the Twin Cities, was a box office smash, as was the soundtrack album. Local music promoters were inviting media from all over to First Avenue, where much of "Purple Rain" had been filmed, encouraging reporters to write about the so-called "Minneapolis Sound" and how it was sweeping the nation. WWTC wanted a piece of the action and the one-time home of the "Golden Rock" would now be the home of the hip local scene that supposedly was the talk of the entire nation.

As the official headquarters of the "Minneapolis Sound", WWTC spotlighted local acts including Sussman Lawrence, the Wallets, Melanie Rosales (whom the station predicted would become as big as Prince), Limited Warranty, and Curtiss A.; real cutting-edge stuff.

Mark Johnson
(Courtesy David Carr)

"We have a lot of local recording here and this station has taken a lot of chances with locals like Prince and Melanie Rosales", operations manager Dick Driscoll told *St. Paul Dispatch* columnist Rick Shefchick. "We're going to do that even more now."

Recalls late-evening announcer Darrel Renstrom, "It was the politics of dancing reflex. Madonna was the flavor of the month, and Cyndi Lauper... Jimmy Jam and Terry Lewis were getting started, and then there was 'Purple Rain'. All that stuff was happening at about the same time."

The air personalities, which were not as important to WWTC's new format as they had been in the past, included Larry Nelson in the morning, followed by Nancy Rosen, Roger Bull in the afternoon and newcomer Benjie McHie, a former evening personality on KQRS-FM (92.5).

The wee-hours were manned by Renstrom from ten p.m. to two a.m. and music director Mark Johnson from two to six a.m. Weekenders Randy Randall, Carol Lasota and Del Roberts (who hated the format so much he refused to use his name on the air) survived the format change. Alan Freed and Steve Amman were added to the part-time staff. It was Freed's second stint at WWTC; Roger Bull had actually fired him several months earlier during the oldies format, but Freed's preference was for contemporary music anyway.

The Trip To Louisville

Nobody at WWTC really wanted to drop the Golden Rock format, least of all Roger Bull and Dick Driscoll. Bull, who came out looking like the bad guy in the eyes of the old 'TC fans, actually got into radio aspiring to work for the Golden Rock. But he was given the responsibility to breathe new life into the station and this is what it came to.

Early in '84, program director Bull and newly-appointed music director Mark Johnson did some research and wrote a proposal on how to take a 5,000-watt AM station and give the Twin Cities market a unique sound. The proposed format was "urban contemporary", which was highly successful in such cities as Detroit, New York and Los Angeles but remained untouched in the Twin Cities except on two low-power community stations. There was a question, however, if such a format would work on AM when urban contemporary was primarily an FM format.

To see just what the potential really was, Bull and operations manager Driscoll took a trip to Louisville, Kentucky where a 5,000-watt AM station just like WWTC was beating out the local CBS station (just like you-know-who in the Twin Cities), playing urban contemporary.

Louisville has a large black community and this particular radio station had a nearly all-black airstaff (except for the overnight jock who was white) and almost entirely played music by black artists.

Bull recalls, "Dick and I sat in on a meeting with the program director, sales manager and general manager and they said 'You're going to make an urban contemporary format go and you don't have any black people working for you? You guys are crazy'!"

The blue-eyed, blond Roger Bull was a little incredulous about the remark. To him, urban contemporary meant music for a city, not music for a specific ethnic culture. He did, however, balance the ticket by hiring the talented veteran broadcaster Benjie McHie.

WWTC RADIO
1280 AM

Tomorrow's Music Today

Top 20
for this Week

1. NIGHTSHIFT
   Commodores
2. COOL OUT/GET IN THE MIX
   Magnum Force
3. PLEASE DON'T GO
   Nayobe
4. BAD TIMES/GOOD TIMES
   Thelma Houston
5. BE YOUR MAN
   Jesse Johnson
6. THIS IS MY NIGHT
   Chaka Khan
7. SMOOTH OPERATOR
   Sade
8. OUTTA THE WORLD
   Ashford & Simpson
9. SMALLTOWN BOY
   Bronski Beat
10. TOTALLY NUDE
    The Wallets
11. ONE NIGHT IN BANGKOK
    Murray Head
12. SENSE OF PURPOSE
    Third World
13. FRESH
    Kool & The Gang
14. DON'T YOU (FORGET ABOUT ME)
    Simple Minds
15. 'TIL MY BABY COMES HOME
    Luther Vandross
16. INNOCENT
    Alexander O'Neal
17. NEW ATTITUDE
    Patti LaBelle
18. BAD HABITS
    Jenny Burton
19. SHOUT/EVERYBODY WANTS...
    Tears For Fears
20. BOY
    Book Of Love

After spending a week in Louisville, monitoring and studying the station, Dick and Roger returned to the Twin Cities and tried to figure out how to make the format work locally.

Driscoll was skeptical. While the black population in Louisville was about 30 percent, it was only about two percent in Minneapolis at the time. Driscoll recalls telling management "if you're gonna use this theory, then we ought to be playing Swedish polkas. . .you might have 30 percent Swedish people in this town but even they don't act in unison."

Driscoll also warned that if the format is successful, and an FM station in the market picks up the format, this AM station will be history — that was exactly what happened to the station in Louisville a few months later.

Instead of being just a "black" station, the WWTC format was formulated to include all kinds of dance music with a heavy beat, especially if it was being recorded locally but was getting little exposure outside the popular night clubs. WWTC aspired to be the home of "The Minneapolis Sound", a concept often talked about by music critics and record promoters in the area.

"The Minneapolis Sound" became popular folklore in the Twin Cities. While Minneapolis has for a long time been a hotbed for rock acts, going at least as far back as the Fendermen, very few local acts have made it big nationally. While local media and others lauded the "Minneapolis Sound" throughout the eighties, claiming that the city was a nationally renowned center for popular music, one would have been hard-pressed to convince music industry people in New York or Los Angeles, in spite of "Purple Rain". But it made "Minneapolitians" feel nice.

On that cold but sunny November day when "Metro Music" replaced the Golden Rock on WWTC, everybody knew there'd be listener resistance. Nobody wanted to answer the studio lines. The front office receptionist required to answer the phone, was bombarded with irate calls throughout the week.

Darrel Renstrom recalls what it was like answering the studio lines. "There were a lot of 'expletive deleteds'. The oldies core was not happy."

When Mark Johnson received complaints during his wee-hour airshift, he responded by asking "Where were you when we had a guy picketing the station all day?" WWTC knew it had to build an entirely new audience to replace the one that was lost.

Name Games

It is almost inevitable that something must go wrong at WWTC. Indeed, as soon as the new format began, problems occurred.

It turned out the name "Metro Music" was a registered trademark for a canned music service. The station received an attorney's letter the first week the format was on the air informing the station of the infringement. The station reluctantly opted for "Metro Radio" instead. Although they wanted to have "music" in the tag line, they figured "Metro Radio" was safe because Brian Short's radio division was called Metropolitan Radio, Inc.

Ironically, Brian Short was pursuing a similar case against a radio station in St. Cloud, Minnesota where a golden oldies station was calling itself "The Golden Rock".

Short, an attorney himself, sent the St. Cloud station his own letter threatening legal action if it continued using the "Golden Rock" name but the St. Cloud station wrote back, telling Short to go right ahead and sue, pointing out that not only was WWTC no longer calling itself "The Golden Rock" but it did not own any copyright on the name or format therefore there would be little to stand on in court. Short decided not to pursue the matter any further.

The Anonymous D.J.

The "Incredibly Ugly" Del Roberts, who was one of the architects of the Golden Rock and a popular WWTC personality in the early years of that format, had returned to the station just before the change after a three-year absence to try and save the sinking ratings. When the dance music format began, Ugly Del agreed to stay on as a weekender at least until he could find a new gig.

"But", he says, "I never used my name [or any name] on the air. I was ashamed to be associated with that format."

Ugly Del hated Metro Radio and off the air was very critical of it. He stuck around, however, playing dance records on weekends for the few extra bucks it earned him. But his lack of interest in the music showed.

One afternoon, Del remembers being given a ten-inch "extended play" record for his shift. The song had a running time of about twelve minutes. He put the record on and several minutes into it, DJ Alan Freed, a dance music enthusiast, burst into the studio, "just livid", as Del remembers.

"What's the matter?" Del asked him.

"You're playing that at the wrong speed!"

It seems the ten-inch record was intended to be played at 45 RPM, not 33 1/3. But Freed seemed to be the only one who noticed; nobody called in to complain.

"The thing that amazed me [about Metro Radio] is that they actually found five or six people who thought that was gonna work", comments Roberts. "We played the Motels; who ever heard of them anymore? We played DEVO, Kid Creole and the Coconuts, Curtie and the Boom Box, Freddie Mercury, I mean we had to play the debris of the record industry."

There were of course, those who listened to and enjoyed Metro Radio. The *Twin Cities Reader* and *City Pages*, the two leading "alternative" weekly newspapers lauded it frequently, with the *Reader's* Frank "Big Ears" Schwartz writing "in a market saturated with limited and unimaginative commercial programming, WWTC [gives] the Twin Cities area the sophisticated sound to match its much ballyhooed public image."

A letter to the editor that appeared in *City Pages* exclaimed "This oldies-turned-urban-contemporary station has done more to enhance the local music scene than WLOL, KDWB and KQRS combined . . . WWTC, by playing new music and local artists, has forced stations like WLOL and KDWB to add records to their playlists faster than they normally would."

The letter concluded "We now have innovative radio in the Twin Cities and that's made radio more enjoyable again."

Former WWTC intern Stuart Held, who by this time was an athletic instructor at the Jewish Community Center in St. Louis Park, recalls the "high-strung, stressed-out clique girls" of the center's aerobics class always tuned to Metro Radio 1280 during their sessions. He remembers the young women would get especially excited when a Madonna song came on.

Mutual Parts Company

Figuring that the trendy "eighties" crowd wouldn't care to have their music interrupted with news on the hour, WWTC dropped Mutual news altogether except during the morning and afternoon drive times to keep listeners informed on their way to and from work. When the news came over the feed at other times, the commercials were recorded and played throughout the rest of the hour so the station could still collect "affiliate compensation", a share of the ad revenue from Mutual.

The Mutual Broadcasting System, reputable for dealing aggressively with its affiliates, abruptly ended its eight-year relationship with WWTC, switching to WDGY (1130) which agreed to carry all the newscasts and Larry King (which 'TC had dropped years before). It was one less income source for the financially-starved WWTC.

With the loss of Mutual network, WWTC added other programming to the station. One of the more unusual programs was something called "The Sofcast Show".

"The Sofcast Show", which ran Sunday mornings at 8:30, featured news and interviews on the subject of personal computers. Each episode also included several seconds of an electronic hiss. When the personal computer owner purchased a translator for $69.95 and hooked it up between the radio and computer, the hiss translated into free uncopyrighted software which included games, educational programs and enhancements for other software programs.

"Sofcast", and a locally-produced companion program called "Computer Line" attempted to attract so-called upward, mobile "yuppies" with incomes of about $35,000 and up to Metro Radio WWTC, whom the station hoped would stay tuned for the trendy "cutting-edge" music it played. Advertisers lusted for the "yuppie" crowd.

Apparently, it didn't work. As Dick Driscoll put it, "Sofcast was a bomb, an absolute bomb".

Still Little Help From Short

While station owner Brian Short wanted WWTC to shed the "golden oldies" image quickly and thoroughly, he was still reluctant to put much money into the station.

The media helped out somewhat. The format change inspired long, positive articles in the *St. Paul Dispatch* and *Twin Cities Reader* and reports on channels 5 and 11.

"It was fun", comments Del Roberts, "watching on Channel 5 Roger Bull looking into the camera and telling the audience that the oldies format was dead in the water, nobody wanted to listen to oldies anymore. . .and that the new format was on the cutting-edge and how wonderful and great things were expected to be and it would be the new direction in which radio would be headed."

Outside the media reports there was only a small amount of paid promotion put into the new format, as was the case in the past.

WWTC did run a rather cheap TV spot which featured bland computer graphics showing the station's letters and a list of bands the station played while Nancy Rosen rhetorically asked off camera, "Why listen to yesterday's music? Today's music is better!"

"We needed money. We had to have money first of all to promote the change in format, secondly to keep promotions going on the radio station or else nobody knows you're there", comments Roger Bull. "But we never got any real support from the ownership. . .all we could get was seven billboards and that was it — seven billboards in the entire Twin Cities."

When the Arbitron ratings came out, Short was devastated — and he seemingly couldn't understand why the new format could bomb so badly. The spring '85 ratings showed urban contemporary had earned WWTC a rating of 0.3; the lowest rating the station saw in its history. Its final oldies rating was 0.8. Brian Short decided to take the most drastic step yet with WWTC, short of shutting it down.

The Ship Is Sinking

On July 25, 1985, Brian Short fired general manager Charlie Loufek after six years of service, replacing him with Sam Sherwood.

Sherwood had quite a track record in the radio business, especially in the Twin Cities. He was among the first disc jockeys at KDWB when that station went on the air in the late fifties and was KD's general manager throughout the sixties, turning that station into one of the most successful rock 'n' roll stations in the upper Midwest. From there he moved on to easy-listening WAYL-FM (93.7) where he developed the concept of advertising radio stations on the sides of milk cartons. While he was general manager at WAYL, Charlie Loufek was sales manager there.

Sherwood had moved to a station in Denver in the late seventies before returning to the Twin Cities. He and Loufek were professional and personal enemies and upon coming to WWTC, there were a few verbal jabs between the two.

As the newly appointed general manager of WWTC, Sherwood had some revolutionary ideas as to what to do with AM 1280.

He called each of the 'TC staffers into his office individually. The first person he called in was music director Mark Johnson. Sherwood praised him as an "artistic missionary" then told him his services would no longer be needed and asked him to leave the premises.

Nancy Rosen recalls being told by Sherwood "You know, you're an asset to this radio station...we're going to change the format and I'd really like you to stay."

"Wonderful! What's the new format?" Rosen asked.

"I can't tell you that at this time", Sherwood replied. Rosen responded by laughing and walking out the door.

Other staffers responded in a similar manner. As Sherwood informed them Metro Radio was headed for the scrap pile, the crew quickly jumped ship, leaving a critical staff shortage.

Alan Freed decided not to go quietly. At the end of his last show he told his listeners of the pending format change, much to the chagrin of Sherwood who wanted to inform the public of the change on his own terms, told of his experience at WWTC as an oldies station as well as urban contemporary and concluded by saying "I've had enough and I'm outta here!" He put on a record, set down the headphones and walked out the door.

Roger Bull telephoned Gary Rawn, a former 'TC oldies DJ who by this time had become program director at KJJO-FM, the current oldies-type station in town. Rawn had already lured over Darrel Renstrom, who took the name "Bryan Thomas" at 104 FM. He had also hired some of the oldies DJs that Bull had fired in the format change.

Roger told Gary what was going on at 'TC and that he didn't know how much longer he'd have a job. Gary told him "Come on over and work with me!" Two weeks after Sherwood took over, Roger Bull defected to KJJO. Others came over as well.

Operations manager Dick Driscoll, who like Charlie Loufek worked with Sam Sherwood at WAYL, stuck around, being the only one who knew the technical aspects of the station, how to operate it on a day-to-day basis and how to meet payroll. At the same time, he was helping the disc jockeys and other staff jump ship, helping them find new jobs and making sure they wouldn't be gypped out of vacation pay and bonuses.

The "Lame Duck" Format

By August, 1985, the plug was pulled on Metro Radio after only nine months. The new format was coming but the new equipment Sherwood ordered was yet to arrive so in the meantime, WWTC began a contemporary hit format which was to last until the new equipment came and Sherwood could launch the new "revolutionary" format.

But the question was, what is the new format going to be? Sherwood, who told the staff and media that something big was about to take place would not tell anyone what it was, not even middle management.

To replace Roger Bull, Sherwood hired 24-year-old Scott Kramer of KDWB AM-FM as program director. Kramer took the position with pride, claiming to be the "youngest P.D. in town".

A whole new airstaff was hired for the temporary format to replace those who left in the mass exodus. Most notable was long-time KQRS personality Dave Dworkin, who had recently been fired in that station's own mid-eighties turmoil, turning up on the afternoon shift on 'TC. Also hired was Beth McCall and former Golden Rock jock Scott Stevens, among others.

Ugly Del Roberts was the one disc jockey who didn't quit when Sherwood took over. Del was loyal for it was Sherwood who gave him his first radio job, at KDWB in the mid-sixties. Del had much more confidence in Sherwood than anyone else working at the station at the time and was excited to work with him again. Del enthusiastically did the morning shift and filled in for anyone else on any other shift.

With the temporary format, 'TC was playing primarily top-ten hits; "Money For Nothing" by Dire Straits and "Dress You Up" by Madonna among them, along with an occasional overly-familiar oldie from the Golden Rock collection, the tapes of which remained in the studio.

As Sherwood jacked around with the local media, dropping clues every so often about the new format, there was much skepticism.

The "alternative" weeklies in particular were skeptical. As critics at the *Reader* and *City Pages* whined over the demise of Metro Radio, nobody really believed any major format change was coming outside the one the station kept insisting was temporary.

"Station management says that what you hear today is not here to stay, but the recent hiring of veteran local radio rocker Dave Dworkin is a pretty good clue to the direction the innovators at 1280 are headed", quipped the *Twin Cities' Reader*. Two weeks later, the *Reader* stated with some smugness, "word on the street is that the beleaguered AM station will stick with a pop music format despite management's protestations that the new format is yet to come."

Surprise. A week later the pop format was gone and so was Dave Dworkin and everyone else. WWTC was now playing, via automation machine, "Good Day Sunshine" by the Beatles, followed by "Raindrops Keep Falling On My Head" by B.J. Thomas, followed by "Good Day Sunshine" followed by "Raindrops Keep Falling On My Head" followed by "Good Day Sunshine". The same two songs 24 hours a day on WWTC. Was this the new format? Listeners were more confused than ever.

6. THE DEATH OF WWTC

All Weather Radio

Some bizarre things were happening at Radio 1280 late in the summer of 1985. Listeners were confused enough when WWTC phased out its pulsating urban dance format for a more moderate pop sound with an entirely new air staff.

Then in mid September, just as everyone was getting used to the new pop format, listeners tuned in to the station and heard just two songs — "Raindrops Keep Falling On My Head" by B.J. Thomas and "Good Day Sunshine" by the Beatles, 24 hours a day. Days went by and listeners went past 1280 on the dial only to hear the same two songs over and over.

Both WWTC and the Minneapolis Police Department received numerous phone calls. What happened at that radio station, people wondered. Was the equipment malfunctioning? Did a disc jockey have a heart attack and die on the air?

The *Minneapolis Star and Tribune* reported that at around three a.m. one morning police came to the studios, on the top floor of thc Wesley Temple Building in downtown Minneapolis to find out what happened. As it turned out, the constant repetition of songs was no accident; the automated selections which made reference to raindrops and sunshine were introducing a new all-weather format on WWTC.

The two selections continued in constant repetition for a full eight days, from September 12-20, 1985. Finally the weather service went on the air.

"The climate is right for a new kind of radio", proclaimed general manager Sam Sherwood in a punned reference to the new format. Claiming that WWTC was the first all-weather commercial radio station in the country, Sherwood told skeptics "If you go outside your house, this is going to be of interest to you. Before you leave your house, before you golf or play tennis or ski, this'll be the thing you'll tune in to before you do anything. . ." He boasted that "this will be almost a lifestyle thing for everybody who lives in our ever-changing climate."

While Sherwood was convinced that music formats were dead on AM radio and that weather radio would spark a new trend in AM programming, he apparently never considered that WWTC, and AM stations in general, are almost impossible to listen to in a thunderstorm with all the static, just when people need to know the weather.

Those who tuned to Weather Radio WWTC heard a very dry, sterile delivery of weather forecasts, temperatures and travelers' advisories by stiff, monotone announcers. Every five minutes, as part of the format, listeners heard a blip. . .blip. . .blip. . .while a prerecorded announcer said "Weather Radio 1280; at the tone, ____ minutes past the hour." This was followed by a soft beep and the weather forecasts continued. There were few commercials as the sales staff was hard-pressed to convince anyone to advertise on this format.

Dick Driscoll was still officially operations manager at WWTC but when the weather format began, he mysteriously disappeared. Nobody was quite sure what had happened to him.

"I wanted to make sure I had all my vacations coming to me because I figured that between Brian Short and the rest that they'd find a way to screw me out of a vacation", he says.

After seeing to it that everyone else received their due vacations and bonuses before leaving, Driscoll went on his own vacation. Upon returning, he quit WWTC for good after an on-and-off career with the station that spanned more than twenty years. "I knew damn well that when I came back I was not about to be involved in the fiasco of all weather radio."

Upon leaving, Driscoll says he put a "curse" on the station. "I vowed when I left there that lo and behold, I said, as we leave 'TC in the hands of all weather radio that it shall never have another ratings point in the book." Driscoll returned to "easy-listening" WAYL-FM (93.7) where he had been heard a decade earlier.

Replacing him as operations manager was Jerry Cunning, who had managed and was a personality on the old KTCR-FM and its successor, "The Cities' 97".

Another WWTC alumnus, Nancy Rosen, turned up in the 10 p.m. to 2 a.m. shift at the station that launched her career, KQRS-FM (92.5). Sherwood tried to coax her to stay with the new format at WWTC to no avail. "When I found out it was going to be all weather, I thought, you're telling me I would be great with the new format? I mean, that was an insult!" she says in retrospect.

Scott Kramer, the self-described "youngest P.D. in town", made a few major blunders in his stint as program director at Weather Radio 1280.

As WWTC got out of the music business, services such as the "WWTC Sound of Music Machine", where a station DJ could be hired to play music at parties and functions, were discontinued.

Apparently Kramer didn't bother telling the White Bear Lake High School Class of '65 reunion about the change. Kramer canceled the Music Machine gig that was rented out for the class reunion without informing anyone involved in it. The twenty-year reunion was left stood up without a music show. Organizers threatened to sue.

In another blunder, Kramer decided late one evening to give listeners a break from the constant parade of weather reports by launching a call-in show called "Talkabout", apparently to find out if anyone was actually listening. Amazingly, *nobody* seemed to be listening. He couldn't get one single call into the studio until another station employee called in to help him out. The two stumbled through the rest of the show pitifully until the weather reports came back on. A few minutes after that, WWTC mysteriously went dead for several hours. Kramer was fired soon afterwards.

It became very clear that weather radio was not catching on with listeners. Before people left their house to golf or play tennis or ski, they were tuning to WCCO Radio for the latest forecast, as they had for years.

It seemed as though Dick Driscoll's "curse" on the station was for real; as Weather Radio, WWTC drew a zero rating. Not only was the listenership less than one-tenth of one percent of the overall listening audience, making it too small to measure, but an Arbitron Company employee told a media critic that they never even saw the call-letters WWTC in any diary the company received from that ratings period. Those call-letters and Weather Radio all together soon disappeared.

KSNE "Sunny 1280"

Sam Sherwood told *St. Paul Pioneer Press-Dispatch* media critic Rick Shefchick that the letters WWTC have "been through so many war-torn adjustments that it's time to wipe the slate clean", adding "they've been abused too much".

By June 20, 1986, Sherwood put the "abused" letters out of misery and the station began identifying itself as KSNE.

In the July, 1986 issue of the *Hornet*, a conservative-leaning "alternative" newspaper, media critic William P. Chadwick eulogized station WWTC, outlining its history "from the 'starburst clusters' of the '70's to an all-news format" and how it later became "the second longest running oldies station in the country".

Chadwick's views on the direction the station was headed were not optimistic. "WWTC was rode hard and put away wet once too often. Owned by a company that had no working knowledge of what the radio world was all about, it never was aimed in the right direction and we can't expect any improvements now." William P. Chadwick was really Del Roberts.

At about the same time that WWTC became KSNE, former Golden Rock personality Steve "Boogie" Bowman, who had more recently moved on to KEEY "K-102" FM, died of cancer. Bowman was considered one of the nice guys in the cruel, cut-throat world of radio. He was and is missed by his colleagues and fans.

Along with the name change for Radio 1280 came new studios and offices. The city bought the Wesley Temple Building from the Shorts, which was demolished to make room for the new Minneapolis Convention Center. The radio station was moved to a much smaller office building owned by the Shorts, located at 215 South 11th Street in downtown Minneapolis, across the street from the Leamington Hotel, which the Shorts owned as well.

Sherwood, meanwhile, signed KSNE to the Satellite Music Network of Dallas, which began feeding a "nostalgia" format to the station on August 4, 1986.

KSNE kicked off its new format on August 1 by spending the entire weekend playing every recorded version of "Stardust" known to exist. The satellite-fed programming took over on Monday, featuring what was billed as "cheek-to-cheek" music by the likes of Andy Williams, Woody Herman, Linda Ronstadt, Julie London, Neil Diamond, Frank Sinatra and Barbara Streisand.

A similar format was originating locally at station KLBB (1400) with live air personalities but Sherwood felt the satellite format would be better because it runs smoother and is far cheaper to put on the air.

As for the all-weather format he was so sure would become a major part of the lifestyles of Minnesotans and would start a nationwide trend, Sherwood told the *Pioneer Press-Dispatch* that he knew it would be a "cume" station, meaning that its ratings would be based on a cumulative audience which tunes in for a minute or two rather than measuring quarter-hour listenership, and that the station couldn't meet the 300,000 to 400,000 cumulative rating needed to insure its survival. He insisted, however, that Weather Radio "could have made it if we'd been able to sustain for another two years. . .but we had to make a business decision."

Billed as "Sunny 1280", KSNE limped along for over a year as a "cheek-to-cheek" nostalgia music station. The format was no more successful than weather radio was. KSNE still wasn't showing anywhere in the ratings while KLBB continued to thrive.

Sam Sherwood, the radio genius who turned KDWB-AM and WAYL-FM into major successes in the Twin Cities couldn't do a thing with Radio 1280. By 1987, Sherwood was gone and Mike McAnally, a former personality at KSTP Radio and Television and sales manager at KQRS, became general manager at KSNE.

K-TWIN Sound Is Back

Station owner Brian Short was disappointed in the performance of the satellite-fed "cheek-to-cheek" music, which he called an "unremarkable middle-of-the-road format". After three rating periods, the audience listening to 1280 was still too small to measure.

Effective November 27, 1987, KSNE dropped the Satellite Music Network and signed on with another satellite service, the Minneapolis-based Progressive Music Network, carrying a format called "The Breeze".

Going from "Sunny" to "Breezy" 1280 (the station seemed fixated on weather themes), KSNE began broadcasting "new age" or "soft fusion" jazz, beamed via satellite all the way from northeast Minneapolis. Music from performers such as Hiroshima, Sade, Earl Klugh, Pat Metheny, Maniheim Steamroller, Jeff Lorber and others was now heard at 1280. This particular style of music is laid back and soothing, the type of music used in "relaxation therapy" tapes.

There were plans to change the call letters to WBRZ but another station held those letters and didn't want to give them up so radio 1280 remained KSNE.

The Breeze had potential for success locally. The format was created by Jack Moore, former general manager and part owner of the defunct Anoka-based radio station KTWN-FM (107.9, also known as "K-Twin"), which had cult following status playing a similar style of music in the late seventies and early eighties.

On the KSNE staff was Brian Turner, who had been a personality on both K-Twin and the Cities '97, and was moonlighting as an announcer on the Breeze network. He was now in the unique position of working for both the network and the local affiliate. Employed as a producer at KSNE, the station had him host a local afternoon show in addition to his duties producing and monitoring the network feed. On weekends, he was heard over the entire network.

“It was kinda freaky”, Turner says. “The 1280 studio was in downtown Minneapolis, the Breeze studio just across the river in northeast Minneapolis, but the signal would travel 23,000 miles up and 23,000 miles back yet be broadcast locally. If I’d listen to myself on 1280, it was about two and a half seconds behind because of [the distance] from the satellite and back.”

Billboards around the Twin Cities announced that “K-Twin sound is back!” at 1280 on the AM dial. But if it once seemed senseless to put urban dance music on an AM-mono signal, it made even less sense to put “new age” music on AM. The music was certainly appealing to young, upscale listeners whom advertisers target, but those young, upscale listeners want to experience the full, stereophonic effect of the music with the expensive equipment they own. Tune in an AM station on that equipment and it sounds little better than a transistor radio.

Once again, when the ratings came out, KSNE-1280 didn’t show up anywhere. Dick Driscoll’s “curse” continued to haunt the station. Of the 23 stations carrying the Breeze nationwide, all but KSNE were FM outlets. The Breeze affiliate in Seattle drew a 4.5 rating compared to KSNE’s 0.0. The format bombed in its own city of origin.

After just six months, the Breeze and KSNE parted company. The ever-frustrated Brian Short decided to scrap the idea of subscribing to satellite-fed formats and scrap the call-letters KSNE all together. In all the years he owned the station, there was only one format that had been successful for him and the time seemed right to bring it back from the dead.

An older and wiser Del Roberts returned to Radio 1280. But you never can really go home again.
(Courtesy Stuart Held)

7. THE PSUEDO-RESSURRECTION OF WWTC

School's Out!

Friday afternoon, May 27, 1988; it was Memorial Day weekend, the unofficial start of summer. As the clock struck three, the beleaguered Twin Cities AM radio station known as KSNE pulled the plug on the satellite-fed fusion jazz music format that had only been heard on the station for six months.

"Ugly" Del Roberts, a voice from the past on Radio 1280 announced "The Golden Rock is back!" Station owner Brian Short ceremoniously pressed a button which started the music, a 15-minute medley of popular rock 'n' roll tunes from 1955 to the present. KSNE was now "The Original 1280 WWTC".

the original 1280 AM
WWTC
Plays the Oldies

"In 15 minutes we'll take you from puberty to high school graduation and your prom, through college. We'll get you in and out of Viet Nam real fast, and take you on your way to being a mom and dad", is how station manager Mike McAnally described the 15-minute rock 'n' roll medley to *Star Tribune* columnist Noel Holston, with visions of stereotypical hippie-turned-yuppie baby boomers in mind.

McAnally told Holston that the popularity of "The Big Chill" movie soundtrack and the use of '50s and '60s music in other movies and in commercials convinced him and station owner Short that now would be the best time to switch to an oldies format. Golden oldies was also the only format WWTC ever had any success with in the ten years the Short family owned the station.

McAnally had approached former WWTC personality Del Roberts, who was doing a Saturday evening program at the Frogtown Diner in St. Paul for KLBB (1400), in early 1988. He told Del of his plans to bring back the Golden Rock to WWTC with all its original splendor and glory and asked him if he would like to be program director.

Roberts, who was by far the most outspoken critic when WWTC dropped the oldies format four years earlier, accepted. This would be his opportunity to not only do it again, but do it "right". Roberts often disagreed with how the various program directors who came and went during the reign of the Golden Rock had handled the format.

Roberts was to be program director and afternoon drive host, while continuing his Saturday night program on KLBB. Once again, the "Incredibly Ugly Del" was back on 1280, telling the same one-line zingers and talking about cruising to Porky's drive-in, as he had in a seemingly earlier time.

Intern Adam Abrams, fresh out of Edina High School and nicknamed by Del as the "Boy Wonder", was Roberts' wise-cracking side kick. They played off each other with kindly jabs as Dick Driscoll and Arne Fogel had so many years before.

The new air staff at the new WWTC was an interesting mix of seasoned broadcasters at low points in their careers and young rookies launching their own careers. Morning drive was headed up by veteran top-40 DJ "Bullet" Bob Lange, a voice at the old KDWB-63 in the 1970s and at WLOL-FM (99.5) through most of the '80s.

The suave Brian "B.T." Turner, held over from the old fusion jazz format, played the oldies weekdays following Lange. Ironically, he also continued to work for the Breeze network on weekends. An alumnis of "smooth-jazz" stations K-Twin and the Cities' 97, he adapted to oldies quite well and displayed a subtle, yet sardonic sense of humor.

Following Ugly Del's afternoon shift came "Sweet" Michael O'Shay, whose resume included the infamous WYOO "U100" FM in the 1970s and more recently KS95-FM and numerous commercials where he provided the voice over.

Originally slated for the 10 p.m. to 2 a.m. slot was a young man in his early twenties who called himself the "G-Man". After a week of fumbling on the air he was fired and replaced by Ray Erick, a rising star who had a short stint co-hosting the morning show on oldies-turned-heavy-metal station KJJO "Hot Rockin' 104". Ousted by the inept managment at that station, he found himself working at a place with even more inept managment.

On the grave yard shift was a promising Brown Institute graduate who called himself "Scooter Pie". One of the more promising of the "green turks" on the staff, he was in his late twenties and had previously worked on the pro wrestling circuit.

On weekends, engineer Archie Stolch was on the air as "Ralph Simpson", along with former KJJO announcers Ray Hartfiel, Bill Hymus and Randy Randall, who was now using the even more schmaltzy name "Rusty Steele". Minnesota News Network provided local and national news updates during daytime hours.

To everyone involved, the all-new WWTC was sure to be a hit. The golden oldies were being dubbed on tape cartridge from compact disc whenever possible and owner Brian Short, known for his tendency to be frugal, expressed a willingness to convert the station to AM stereo by the beginning of 1990. Certainly, it was anticipated, the station should at least attract a large enough audience to actually draw a rating, something it hadn't been able to do in three years.

But Mike McAnally, the current general manager, was no Charlie Loufek and operations manager Jerry Cunning was no Dick Driscoll. Times had changed considerably in the past few years and the things that made the Golden Rock work in the early part of the decade no longer existed within
the station or within the local radio scene. Oldies had gone Hollywood and a 5,000-watt AM station just wasn't enough anymore.

Ray Erick
(Courtesy Del Roberts)

A Long, Hot Summer

The very first week the all-new WWTC was on the air, there was already an incredible amount of internal turmoil. Everybody had their own idea of what direction the station should go in and as a result, little was accomplished.

Program director Roberts wanted the station to be the spontaneous free-for-all it was in its glory days, with listener calls put live on the air to make dedications and tell jokes, disc jockeys walking in on one another's shows and a music library that included almost anything and everything that was a hit, from the roots of rock 'n' roll all the way up to the present. General manager McAnally disagreed.

As the Ugly Del recalls, "McAnally came to me over at the Frogtown Diner, said he wanted to hire me, wanted me to be p.d. and his exact words were: 'we want you to do exactly what was done before in every way, shape, manner or form, recreate the excitement of the format of the old 'TC'. Second day I was there he said we ain't gonna do the old 'TC."

McAnally, who understood current trends more then sentimentality, wanted the spirit that inspired listener loyalty and advertising dollars in the old days, but adapted to the times. There was no getting around it; the marketplace had changed. In the time that had passed since the glory days of the Golden Rock, MTV and cable television had revolutionized the culture tremendously, changing the scope of how people treated the radio medium as a whole, let alone a two-bit AM station like WWTC. There was no going home again.

With visions of the all important baby-boomers with incomes of over $30,000 on his mind, McAnally reminded Roberts that he was the boss and insisted on doing a much more structured format of just big hits, not too much chit-chat from the announcers and no instrumentals, for fear listeners would get bored and tune out during songs they couldn't sing along with. He reasoned that yuppies were much too serious-minded for a lot of dumb jokes and that no tunes released past 1974 should be played because, after all, we're playing baby-boomer music!

Says Brian Turner, "You're not going to 'bring back' anything. Radio is a

24-hour-a-day, 7-day-a-week, foreward-moving medium. You can't think backward. You can look backward to learn some lessons and maybe grab some things that you can take with you but you can't live back there."

In what seemed to be a particularily dubious programming decision, the station had made a deal with KARE-TV (Channel 11) to simulcast that station's five p.m. newscast in exchange for cross-promotion (but no money) between the two stations. This cut directly into Ugly Del's airshift, forcing him to stop the music for a half hour every day, coming back on at 5:30 for another half hour, turning over the microphone to Sweet Michael O'Shea at six.

The music stopped again Thursday nights from seven to eight for a fluffy, innocuous talk show called "Twin City Beat", hosted by Marcia Appel and Bill Dorn, more or less a radio version of the decidedly dull *Skyway News*.

Roberts had been hired as program director but the title seemed to be in name only. McAnally insisted on being involved in every aspect of the station's day-to-day operation for better or worse. After less than two weeks and several heated arguments with McAnally, a frustrated Roberts resigned as program director, agreeing to stay on as an afternoon disc jockey. The program director title was handed over to Bob Lange.

Lange was decidedly less of a troublemaker than Roberts. The fiercely strong willed Roberts came from the old school of rock 'n' roll radio where the DJ is the master of ceremonies, entertaining the listener with quick bits of humor along with information such as time and temperature, with the music thrown in as almost a secondary thing, in a fast-paced format.

McAnally and Lange came from the more contemporary school of FM radio, where the philosophy is more emphasis on music and less on the guy who's playing the records. Lange at least attempted to see both sides of the spectrum.

"I tried to run the station with no budget, I tried to have fun", Lange says. "WWTC had inherent problems and the whole philosophy of mine was to run the station and give people chances to be on the air who had never been on before or had always wanted to do an oldies format. [There was] no budget, so it was a challenge."

Lange, who wasn't really fond of having to be on the air at six in the morning, switched to the more tolerable noon-three p.m. shift. Michael O'Shay took over on the morning drive and Ray Erick was moved to early evenings, from seven to ten p.m. Sensing talent in the 18-year-old "boy wonder" intern named Adam Abrams, Lange began to give him weekend air shifts, which McAnally was initially against.

Recalls Abrams, "I was 18, I was happy to be there, everything was cool...as far as a place to start, there couldn't have been a better situation. There were guys who knew what the hell they were doing and probably shouldn't have been there. I'm on the air with them with *no idea* what the hell *I'm* doing."

"The Original 1280 WWTC" came to be, for the most part, little more than an abyss of mediocrity. Lange tried to be open-minded, but McAnally was known, among other things, to pull songs from the music library for fear they might offend someone, including "Hi Hi Hi" by Wings and "Indian Giver" by the 1910 Fruitgum Co. The station even played an altered version of Lou Christie's 1966 hit "Rhapsody In the Rain", where the line "making out in the rain" was replaced by "fell in love in the rain" and "our love went too much too far" was replaced by "love came like a falling star".

Adam Abrams, the Boy Wonder.
(Courtesy Del Roberts)

Even after resigning as program director, Ugly Del continued to get under McAnally's skin when he did his afternoon show. While McAnally was very cautious about not putting anything over the air that might offend someone, Del often and without apology, made off-color jokes and puns during his show. He also talked more than McAnally would have liked, although the half-hour reserved for KARE-TV news was all talk anyway. He apparently didn't consider that people were more likely to tune out then.

The older members of the air staff sat back, put in there time and went home, not letting the internal turmoil bother them. After all, they were allowed more freedom than they would at a station that actually had ratings but the younger, more idealistic ones had high hopes for the station's potential and were quickly disappointed.

Ray Erick, who fondly remembered listening to Del, Nancy Rosen and B.J. Crocker in high school and got into the business in part because of the Golden Rock, was hired by Roberts and the two had talked of him possibly being music director or perhaps assistant program director.

Erick had been assembling the bulk of the station's music library, converting tunes from his own collection of compact discs to tape cartridge when McAnally decided to hire a young record collector named Stuart Held as the official music director. The 25-year-old Held was an intern with WWTC when he was in high school, during the original Golden Rock period. He had since done odd jobs at several area radio stations, and had worked in college radio, but had no commercial on-air experience.

There was a lot of tension between Held and Erick early on; Erick, a fairly well-known local talent, had been working for that title and felt Held wasn't qualified for the job.

"I carted up the first 300 records on that station when it went back to oldies out of my personal collection", he recalls. "99 percent of them were [from] C.D. and we would run them through the equalizer and make them sound really good which made that AM station sound nice enough where it sounded like an inferior FM. Well, then comes Stuart.

"Stuart comes in with a stack of old records that he dug out of his basement and started dubbing solely off album, stuff that I could get on C.D., bypassing the equalizer. Stuart's dubs always sounded different."

Held admits that he had no idea how to run a C.D. player when he came to the station and was unfamiliar with a lot of the equipment. While he and

Erick eventually set aside their differences, Held quickly found himself involuntarily playing the role of whipping boy for a lot of the hotter egos around the station.

"People there weren't treating me all that great, especially that first week", Held recalls. "[Operations manager] Jerry Cunning was like, 'who are *you*? Are *you* supposed to be here?'" Held also recalls being chewed out mercilessly by Michael O'Shea because he dubbed the song "Twenty Flight Rock" by Eddie Cochran and O'Shay felt it wasn't a big enough hit.

The bad mood and attitude problem a lot of people seemed to have at WWTC even came out on members of the public who were simply interested in the station, people who in happier times were known affectionately by WWTC staffers as "groupies".

When "The Original 1280" did a remote broadcast from St. Paul's "Taste of Minnesota" festival in July 1988, only one listener came by specifically to meet the WWTC DJs. Nighttime jocks Ray Erick and Scooter Pie were broadcasting from a recreational vehicle parked at the festival, while operations manager Jerry Cunning, a large, older fellow, engineered.

The two disc jockeys were cordial with the fan who showed up, making him feel welcome. But when the fan peaked inside the recreational vehicle, curious to see the set up, operations manager Cunning shouted *"This motor home is for authorized personnel* **ONLY! GET OUT!!**" The now former WWTC fan quickly disappeared.

But morale at the station wasn't entirely negative. Says Brian Turner, "I had a real good time there, so I guess I was one of the happy-go-luckies. We had some fun people there, we did some events right off the bat, doing things in the community as best as we could, it being the small, understaffed, underpaid organisation that it was. If there was any explanation for 'the grumpiness' it might have come from the fact that everybody knew this was not going to be the old WWTC. There's no way it could be. But there was still some fun to be had there so I certainly enjoyed it.

The WWTC motor home: Do not enter unless you're authorized personnel! **(Courtesy Del Roberts)**

The Magical Mystery Tour

One of the few truly innovative programs on the new WWTC was a Sunday night feature called "The Magical Mystery Tour", a three-hour excursion into psychedelic music, poetry and nostalgia.

The "tour" began at 10 p.m. with the playing of the Beatles song of the same name. Host Ray Erick would then welcome the "passengers" (i.e. listeners) to his "magic bus" with a voice mellower than usual, mimicking the old "underground" FM announcers of the late '60s and early '70s. Joining him in the studio was his spaced-out partner Mike Velin.

The "Magical Mystery Tour" program was conceived by Erick and Ugly Del. As Erick recalls, "Del and I were sitting around saying, 'you know, wouldn't it be fun to have a nightclub that plays the Doors, the Jefferson Airplane and the heavy stuff that the Heads would know, a nightclub where everyone sits on the floor on pillows and a hint of pot is in the air. . .

“I said that would be a good idea for a radio station and he said ‘at least a show on a radio station’”. Inspiration hit Ray and Del simultaneously and the Tour premiered in mid-June of ‘88.

The music on the Tour consisted of mostly wild, sometimes obscure psychedelic album cuts and B-sides with an occasional familiar tune thrown in. Between the songs, Ray and partner Mike would speak of good vibes and free love with spontaneous silliness, while Ravi Shankar sitar music played as a backdrop. Ray would sometimes read Alan Ginsberg poetry while his partner created gurgling sound effects in the background, suggesting he was firing up a “bong”.

The program quickly found its own cult following. “We had some characters that kept [calling in] on that show as regulars and we thought if they weren’t [going to be] on the show, they should have called in sick and told us”, Ray remembers. The regular “passengers” on the “bus” included Dennis and his wife Cindy, who offered up a mantra every week, Dr. Timothy Treeman, a spacey poet who recited his own works on the air and three young college girls named Margro, Ana Lisa and Kelly who would offer their own perspective of the “trip”.

While doing the “Magical Mystery Tour” program, Ray and Mike fell into the mood by decorating the studio with a psychedelic atmosphere. Colored lights and lava lamps were brought in and set up in the broadcast booth every Sunday night and all the regular lighting was turned off while they did the show. Incense was burned and the hosts wore paisley shirts, bell bottoms and sandals while doing the program, even though it was only radio.

General manager Mike McAnally walked in on the “party” one night and gasped at what he saw. Upon seeing something smoldering in the broadcast booth, among the flashing lights, he pointed and said “You’re not allowed to smoke in the studio!”

Erick, a non-smoker, assured him “I’m not smoking — the incense is.”

There were objections to the program; McAnally and Bob Lange cringed whenever subtle references to drugs were made (on one show, for example, Ray asked his “passengers” for their favorite mushroom recipe) and

McAnally questioned Erick's qualifications for doing such a program; at the age of 22, Ray was too young to be a baby boomer.

Management put up with the Tour, at least for the time being, figuring nobody was listening after 10 p.m. on a Sunday night anyway

I Did Not Apply At 93.7 FM

When WWTC decided to play oldies again, the only other oldies station in town was the moribund KDWB-AM (630) whose programming was beamed in via satellite from California. It would be a cinch to beat that station out, those at 'TC felt, with locally originating programming. But "K-63" didn't stop the music for TV news simulcasts or talk shows and while that station had consistently low ratings, it was at least showing up in the Arbitron books.

Adding to the woes of WWTC, almost as soon as the station switched to golden oldies, two FM stations picked up the format. WAYL-FM (93.7) dropped its longtime "easy listening" format in favor of "classic hits" on July 22, 1988, just two months after 'TC went oldies, changing call-letters to KLXK. KMGK-FM (107.9) followed suit on October 1, becoming KQQL "KOOL 108" FM.

In staff meetings, program director Lange assured that the new competition was no major threat to WWTC, things were just fine and there was nothing to worry about. But Lange and other staffers were looking for greener pastures.

An acquaintance of Lange who was working for KLXK had contacted him about employment at the new FM oldies station. Lange came down and talked, expressing interest but telling him he was committed to WWTC. On his way out, he happened to cross trails with Brian Turner. The two men looked the other way, pretending not to notice each other.

A woman who worked for KLXK related the incident to a friend at WWTC. The word spread to the proudly irreverent Del Roberts, who came to work the very next day wearing a T-shirt that said "I DID NOT APPLY

AT 93.7 FM". Lange and Turner both saw him that day and were not amused, much to his delight.

A few weeks later, when it looked as though KLXK wasn't going to hire either Lange or Turner, Del had another shirt made that said "I WAS NOT REJECTED BY 93.7 FM".

Listen Closely And You'll Hear the Ratings Drop

When Ugly Del told a corny joke during his afternoon program on WWTC, he would often quip "Listen closely and you'll hear the ratings drop".

In actuality, the ratings had nowhere to drop. The station had, according to Arbitron, less than one-tenth of one percent of local listenership since switching to an all-weather format in 1985. Subsequent formats did no better for the station.

When the Summer 1988 Arbitron ratings came out, it was the same old story. Despite high hopes, the new golden oldies format was doing no better than Weather Radio; still a rating of 0.0. Driscoll's Curse lingered on. Station ownership, however, expressed a supposed commitment to the new format, accepting that it might take a while for the station to make a complete turnaround. But it was clear WWTC was far from being the "big-time" station it once was.

Bob Lange, who had spent years at some of the highest rated top-40 stations in town, had to deal with the lack of ratings, the lack of morale and an ownership that was unwilling to spend anything to keep the station afloat or knew anything about how to run such a business.

"You had to let everybody do what they wanted to do because you couldn't pay them a damn thing", Lange says. "They could [practically] make more money working for four hours at a Burger King drive-thru than they could working a week at WWTC."

Brian Turner, who enjoyed working with the format, nevertheless jumped ship to KJJO-FM in September for the prospect of better pay and his own morning show. The owners didn't want to finance another full time staff member so with Turner's departure, everyone's air shifts were expanded. The ten p.m. to two a.m. shift had also been eleminated so this meant that the two nighttime personalities, Ray Erick and Scooter Pie, now had to pull off six-hour shifts, with Erick on from six to midnight and Scooter on from midnight to six in the morning. Erick, who had his share of differences with Lange, was beginning to suspect that he was being pushed out the door.

"The shit hit the fan", as Erick puts it, in early November when he came in for his evening airshift and noticed Lange's car in the parking lot. The program director is usually gone for the day before six p.m. When Ray sat down in the broadcast booth, plugged in the headphones and was about to open the mike, Lange stuck his head into the door and said "Segue into another record".

Lange walked into the studio with his own stuff and told Erick he had to let him go. When Erick asked him why, Lange refused to tell him until Erick pressed on and Lange merely said, "it's your attitude", refusing to be more specific. Ray was handed a written notice of termination the following Monday by operations manager Jerry Cunning, who, according to Erick, literally threw the notice at him and yelled "You have to leave! You can't even be in the building!"

Says Erick in recalling the incident, "I thought, what is with these people? Does it look like I'm carrying a machine gun, do I look like a terrorist? What am I gonna do? I want my damn letter of dismissal!"

Before exiting WWTC for good, Ray went into the production studio and destroyed all the tapes of music used in his "Magical Mystery Tour" program with an electronic demagnetizer.

"There was no way I was gonna let them rape that show and let somebody else do it", he says with some bitterness lingering. "I volunteered all of my time when I put all those songs together and taped everything. I figured they were not getting all this free stuff from me."

Fired from the AM station with the consistent zero rating, Erick became the evening DJ at "KOOL 108", the FM oldies station, where he found himself far more appreciated. He later went to the Twin Cities' top-rated FM station, KQRS, where he often filled in for the exceedingly popular Tom Barnard and the KQ Morning Crew. Bob Lange showed him!

With Erick's departure, late night DJ Scooter Pie moved to early evenings. Engineer and occasional weekend jock Ralph Simpson took over on the overnight shift for a short time, replaced in December by a woman who called herself Alix Kendall. Kendall showed a lack of knowledge about the music she played from her first night on the air; she played a song by the Chordettes, pronouncing the 'ch' in the group's name. She nevertheless was quickly taken off the graveyard shift and moved to late mornings, nine to noon.

As 1988 came to a close, a burned out Del Roberts left WWTC for good, taking a job as program director at an oldies station in Sioux Falls, South Dakota, where the station manager was actually willing to allow him some say in how the format should be run. He was replaced by Dick Ervasti, a broadcast veteran who, like Lange and Michael O'Shay, was well known for voice overs in commercials.

Songs From The Basement

With Erick's departure in November, the Sunday night time slot which his "Magical Mystery Tour" program had occupied was given to an even more creative eccentric, music director Stuart Held.

Bob Lange, who didn't seem so enthusiastic about Erick's creative ideas, suggested to Held that he fill the Sunday night slot with his own "format" show, something different from what Ray had been doing. On November 13, 1988, "Songs From the Basement" premiered on WWTC.

"Songs From the Basement" was a cavalcade of album cuts, B-sides and often a bizarre obscure song, all from Stuart's personal record collection which he kept in the basement of his home.

For diversion he brought in his brother Joel as his on-air partner and "straight man". The Held Brothers casually talked trivia and often got silly as well, saying whatever went through their minds at the moment.

The "Songs From the Basement" program sounded something like a show on a college radio station or even an illegal "underground" station run by a couple of high school kids; the Helds were known to do some strange things on their show. One week they played an obscure 1910 Fruitgum Co. record called "Pow Wow" backward, which revealed the message "Bring Back Howdy Doody". In another installment, they played part of a defective David Cassidy record in which the hole for the spindle was off center, making the record sound warped. There was, of course, plenty of enjoyable and interesting music on the program as well.

The Helds even did a tribute to the original Golden Rock one evening, playing tapes of WWTC broadcasts featuring the likes of Steve "Boogie" Bowman, B.J. Crocker, and Dick Driscoll & Arne Fogel doing "Stump the Chump".

Stuart Held enjoys a snack in his basement.
(Author's Collection)

The night before "Songs From the Basement" made its debut, Stuart Held says he lost sleep, not so much because he was nervous about going on the air, but because "I was worried about Ray Erick. It was really, like, this is Ray's show." Stuart was worried that "Magical Mystery Tour" fans would call up in droves and cuss him out and make threats. To defuse any possible tension, he made a point of acknowledging Ray on his first show and wished him well in his pursuits. Old fans of "The Tour" did call in, but only out of curiosity. Many, including the "Tour's" resident poet, Dr. Timothy Treeman, complemented the brothers and promised to keep on listening.

Much to Stuart's relief, he was able to win over much of Ray's cult following and received no particularly negative calls.

Stop The Music

By the spring of 1989, the station that was in perennial dire straits was going nowhere faster than ever. The two FM oldies stations were holding it down to be sure but other factors, such as owners who wouldn't finance the opperation or do anything about the inept management team also contributed to the station's stagnation.

As afternoon DJ Dick Ervasti left after only five months on the air, the station hired Barry Ze Van, the popular TV weatherman, to play the oldies from two to five p.m.

Ze Van, who was remembered for his humorous weather forecasts on KSTP-TV (Channel 5) in the early '70s and WTCN/KARE-TV (Channel 11) in the mid '80s was brought on board primarily for his name recognition but he left something to be desired as a rock 'n' roll disc jockey. He bumbled and stumbled, which was part of his appeal when he did the weather on TV but it didn't sound right on a fast-paced rock station.

In addition to putting Ze Van on the air, WWTC attempted to upgrade its profile by picking up hourly news and occasional sports programming from the NBC Radio Network beginning in April, 1989.

The attempts to generate revenue for the station began to look more and more desperate as WWTC began to forego musical programming for some rather awakwardly-fitting talk shows.

They were still simulcasting KARE-TV news from 5-5:30 p.m. which didn't earn the station any money from KARE, only one ten-second spot on Channel 11 each day at 4:45 p.m. which only told viewers "KARE-11 news at five can be heard on WWTC-1280." 'TC, on the other hand, promoted the simulcast heavily throughout the day, as if anyone would actually be inclined to tune in.

With music listeners tuning out in droves at five p.m., the station added "Ray Scott Sports Talk" at 5:30, meaning the music would be off for a full hour and a half every afternoon.

The music also stopped from nine to ten a.m. weekdays to make room for the "Golden Opportunity Auction", which was more or less a radio version of a "shop-at-home" TV program where listeners could call in and "bid" on merchandise and entertainment tickets.

The auction program did generate money for the station as it was able to keep the profits from what was sold. If the two or three people who were actually listening called in, the show almost paid for itself.

The music stayed off for yet another hour on Fridays right after the "Auction" for a show called "Thanks a Million", a shameless exercise in poor taste hosted by Edina-based philanthropist/charlatan Percy Ross, who gave money to the listener who called in with the best sob story. Ross, a controversial local millionaire who also wrote a syndicated newspaper column with a similar theme, publicly vowed to give away his entire fortune before he dies.

"Thanks A Million" did gain WWTC some national attention. In the winter of 1989, the ABC News program "20/20" did a profile of Percy Ross which included footage of him doing his radio show, the WWTC call-letters clearly displayed on the microphone. Unfortunatly the station had already dropped the Ross program by the time the segment aired.

Weekends brought listeners a Saturday morning call-in show on gardening, a pro wrestling talk show, Notre Dame football and whatever other play-by-play the NBC Radio Network had to offer.

Common sense would suggest to anyone with even a minimal knowlege of radio that block programming almost never works in a music format. Research from all over the country had already shown that those who tune in for music are going to go elsewhere when the music stops and not come back, especially with two FM oldies stations on the dial.

It would have almost made sense to either switch to an all-auction format or generate revenues by doing a better oldies format and putting promotion into it. But logic and common sense, unfortunatly, evaded the minds of those in charge of running WWTC.

Ryan's On the Air Again

Mike Ryan made his first appearance on WWTC in more than five years on May 8, 1989 as a temporary replacement for Bob Lange, who had gone into the hospital for major surgery.

In the years since leaving 'TC in the height of the original Golden Rock format, Ryan had worked locally for KJJO-FM (104.1), WLTE-FM (102.9) under an assumed name, the now-defunct KRSI (950) and for a tiny AM station in Stillwater using the old WTCN call-letters. He had a gig spinning records for a Holiday Inn nightclub in St. Paul when he agreed to fill in on Lange's noon to 3 p.m. shift.

For listeners who yearned for the good old days of 1280 AM, the friendly voice of "Ol' Records Ryan Himself" was a ray of light. He was using the same schtick he used in the past, including a new, rather silly theme song sung to the tune of "Happy Days Are Here Again".

"Ryan's on the air again/Mike Ryan's on the air again/let's sing a song of cheer again/Ryan's on the air again. . ."

When Lange returned a few weeks later, he and Mike McAnally decided to give Ryan a permanent spot, giving him the 2 to 5 p.m. shift occupied by Barry ZeVan.

"That was tough because Barry was a great guy", Ryan says. "It had been a long time since he was in radio and they were asking him to do something he wasn't comfortable doing. When you're trying to put a square peg in a round hole, it's difficult."

Just weeks after returning to the air on WWTC, Lange jumped ship to "classic hits" station KLXK-FM in July 1989. Lange's on-air replacement was Brian Turner, who returned to WWTC after ten months of working for KJJO-FM. Turner had left for the prospect of better pay and better exposure but found himself unhappy at a station that had gone through at least as many format changes and personnel turnover as 'TC had over the years.

The title of program director was passed on to Ryan. He had been PD at the Golden Rock back in 1981 and he, like his old associate Del Roberts, thought he could bring WWTC back to its grand glory.

He merrily added new selections to the music library and pulled what he considered "yucky '70s" music. Music director Stuart Held, who worked with Ryan at the old 'TC as an intern and agreed with his programming ideas, warned him Big Boss McAnally might object to his plans to "bring back" the old 'TC. Stuart had felt McAnally's wrath concerning what music he wanted and didn't want on the station too many times before, as did Ugly Del.

Sure enough, Ryan was called into McAnally's office, where a stack of cartridges Ryan had dubbed sat on his desk. McAnally went through the stack. "Did I approve this? Did I approve this? Did I approve that?" McAnally told him the rules, that any music that goes into the station's library must be approved by him first; nothing that didn't make *Billboard's* Top-20 and no instrumentals either. As Del had learned before, the program director didn't necessarily have the right to program under Mike McAnally.

During Ryan's stint as program director, some new talent was brought in, including Robin Bankes, a female voice to replace Alix Kendall, who followed Bob Lange over to KLXK. Weekend benchwarmers included Domino, an old album rock personality who had worked at KQRS and KDWB-FM when it was known to local "metal heads" as "Stereo 101", Michael Jaye, who had been a news photographer at KMSP-TV (Channel 9)

Mike Ryan, Ruth the receptionist and Robin Bankes play Peek-A-Boo. **(Courtesy Stuart Held)**

and Stix Franklin, who had done some off the air work for the station before being fired due to budget cuts. He was rehired as a weekend air personality under a new budget.

Ryan tried to remain optimistic at the refurbished Golden Rock but it was difficult not only because he wasn't allowed to do what he wanted, but also the morale of the staff remained low. In December 1989, he resigned as program director but remained as an on-air personality, just as Ugly Del had done a year and a half earlier.

The last straw in his decision to resign as program manager came when he was forced by McAnally to fire Stuart Held. Ryan and Held were friends and Ryan didn't see any problems in his performance at the station. But McAnally decided he didn't want Held around anymore and the man who insisted on doing every other program directors' duty himself demanded his program director have the unenviable task of axeman. Michael O'Shea replaced Ryan as program director but it really made little difference; "The Original 1280 WWTC" was already a sunken ship by this time.

1280 Going Once...Going Twice...

WWTC Radio was part of the estate left to Brian Short by his late father, Robert. In the years since his father's death, Brian dismantled much of the estate, selling off the assets. Most of the downtown Minneapolis landmarks his father owned fell to the wrecking ball to make way for parking lots and ramps. A large percentage of the parking space currently in downtown Minneapolis is owned by Brian Short.

In the meantime, the moribund AM station fell to neglect as the owner dealt with other businesses in the estate. Since inheriting WWTC in late 1982, Short lost millions of dollars on the station that had hit rock bottom and just couldn't bounce back. He had difficulty selling it as well, as there was little interest in AM radio in the 1980s and the offers he did receive were for far less than what he felt the station was worth.

Finally in early 1990, it was officially announced that station WWTC would be sold to Edina-based C.D. Broadcasting Corporation, headed by Christopher T. Dahl. The selling price was $950,000; by comparison, Minnesota Public Radio agreed to buy WLOL-FM later that same year for $11 million.

C.D. Broadcasting owned stations throughout Minnesota and the Dakotas, mostly in small markets such as Bemidji, Minot, Redwood Falls, Crookston, Grand Forks and Aberdeen, plus two stations in Maui.

Mike Ryan, who still holds some feelings of sentiment toward the station and format that had been a big part of his career remembers feeling a little saddened over what appeared to be the imminent demise of the Golden Rock and possibly the WWTC call-letters as well.

"Rumors of the sale were rampant and after I resigned as program director and it was like waiting for the other shoe to fall...it was real tough because you knew [it] was something that will probably never happen again."

Ironically, at the same time the sale was announced, WWTC won rights to carry Minnesota Gophers men's basketball and hockey for the next three years. The new owners were not interested in the Gophers' broadcast franchise, however, because the format they intended to introduce to WWTC was to be geared not to adults but to very young children.

Kiddie Radio

"I remember the day that they announced officially that the station had been sold", recalls weekend DJ Michael Jaye, "and the FCC application did indeed say that the proposed programming was to be 'children's educational/ entertainment'. . .I remember we stood in this back engineering room and literally laughed so hard we were crying. We were all to be replaced by Big Bird, Ernie and Bert. There was a lot of skepticism among the airstaff about this thing."

WWTC's next incarnation was to be something called "The Educational, Sensational Radio AAHS". It was no joke; the new owners were not only dead-serious about putting children's programming on WWTC but they planned on making it the flagship of a nation-wide satellite network of children's radio stations.

Although the programming was to emphasize education and somewhat resemble the kind of children's programming seen on public television, the new format was to be a commercial operation. While ratings service companies don't count listeners under age 12, research conducted by the new investors found that pre-teens purchase and consume large amounts of fast foods, toys, music, sporting goods, bicycles and convenience store goods. The under-12 crowd seemed to be a very good, largely untapped market for advertisers.

While the station awaited federal approval of the ownership transfer in early 1990, "The Original 1280 WWTC" sounded very much like a slowly dying radio station.

Cutting operational costs down to the bone, the station laid off much of its staff, both on and off the air. Mike Ryan and Robin Bankes departed as well as some weekend people and the remaining DJs had their hours extended. Evening jock Scooter Pie jumped ship to WLOL where he took the name Adam Savage and remained until Minnesota Public Radio took over that station in February 1991. Adam Abrams, the kid who started as an intern with the station fresh out of high school and later did some late night shifts on weekends, was hired over to KOOL-108, where he soon became the assistant program director and one of that station's most popular personalities. Also as a cost-cutting measure, WWTC began signing off from midnight to 6 a.m. in its last few months.

The FCC approved the sale in May 1990 and the new owners announced the children's programming format would be launched at noon Saturday, May 12 with a live remote broadcast from Como Park in St. Paul. Meanwhile, as WWTC continued to play the oldies, airstaff was ordered to identify the station as "The Educational Sensational Radio AAHS", mentioning the WWTC call-letters only at the top of the hour.

The last day of the old format was a hectic one at the WWTC studio at South 11th Street in downtown Minneapolis. Among the preparations for the new format, the new management had the task of telling the incredulous officials of the NBC Radio Network and producers of Ray Scott Sports Talk that their programming was being abruptly canceled after today. They even finally dropped thc KARE news simulcast.

A few scavengers came by the station on that last day, snatching up souvenirs of the old WWTC, such as stationary, old pictures, banners from station-sponsored events and the WWTC lobby signs. Ugly Del Roberts was one of the scavengers, as was a local radio historian who told bemused staffers he planned on writing an book on the ill-fated story of WWTC.

Throughout the last day and into the next morning before the format change, WWTC listeners became acquainted with the new "Radio AAHS" format with promotional spots advertising programs with titles such as "The All-American Alarm Clock" and "Alphabet Soup". The ads included a cutesy jingle in which a chirpy female voice sang "We have magic for you every day/Just tune to Radio AAHS and let it play/There's lots of fun stuff you can learn and do/'Cause kids are the people that we're talking to. . ."

Listeners were invited to the Radio AAHS inaugural at Como Park Saturday afternoon where they could meet station personalities and load up on free pizza, soda pop and balloons.

WWTC personality Brian Turner remained with the new format as program director and host of the early-morning "All-American Alarm Clock" program. Michael Jaye and Stix Franklin also stuck around and the station rehired, of all people, Mike "Records" Ryan as a weekend disc jockey.

Effective noon, May 12, 1990, the Golden Rock died its second and final death, but one thing would remain the same; the station would still be officially known as WWTC.

W W T C

1 2 8 0 A M

8. THE RADIO AAHS PHENOMENON

Live From Como Zoo

When WWTC weekend personality Domino segued from "Love Me Two Times" by the Doors to a Sesame Street disco record at 12 noon on Saturday, May 12, 1990, the rumors were confirmed; WWTC really was switching to a kiddie music format.

AM 1280 had become "The Educational, Sensational Radio AAHS", a program format "for children and their families". The dried-up, has-been radio station known as WWTC became the launching pad for what the new owners hoped would become a nation wide network of kid-oriented radio stations.

The inauguration of Radio AAHS was celebrated at the Como zoo park in St. Paul that Saturday afternoon. As the happy-go-lucky sounds of children's music filled the air, dozens of kids and their parents, mostly white, suburban, baby-boomer moms, roamed the park, taking advantage of the free Domino's pizza, soda-pop and balloons. The "Noid", a Domino's pizza mascot with large ears and a bright red body, made a personal appearance as did performing clowns and a young "princess" who made face paintings for the children.

Brian Turner hams it up at Como Park.
(Authors Collection)

Radio AAHS posters, featuring a colorful unicorn logo, were handed out, along with a newsletter announcing the "birth" of Radio AAHS and the proposed Children's Radio Network.

"One of the most exciting events in recent radio history is taking place right

Clowning around at the Radio AAHS inaugeral party
(Authors Collection)

here in the Twin Cities!" read a message from station owner Chris Dahl in the newsletter. "...our newly acquired radio station, WWTC, will begin a full schedule of programming specifically for children 12 and under and their parents. This is the first station of what we hope will be a national network of stations serving the listening needs of our nation's preteens."

A stage was set up and Radio AAHS personalities Brian Turner, Stix Franklin and Robin Blair spoke and performed. The new owners of WWTC, Chris Dahl (of C.D. Broadcasting) and Bill Osewalt (of Children's Radio Network), made gung-ho speeches about what Radio AAHS was all about and what its goals were.

The crowd applauded as the funky Radio AAHS music came through the loudspeakers, marking the initiation of the new format. The small children around the park danced to the music.

Soon Robin Blair, a newly hired station personality with long, blonde hair and and a sweet voice, came on stage and performed , live on the air, her rendition of the classic Bobby Day song "Rockin' Robin".

The party at Como Park was a big success. Publicity was drummed up for the new format and the new owners were geared up to begin an aggressive public relations blitz.

People involved in education and child rearing had no shortage of praise for this new concept in radio but radio industry people and others thought it was a rather grandiose joke.

Working Out the Bugs

Quite a variety of programming was offered on this curious new format. As the day began, "The All-American Alarm Clock", with Brian Turner (and later Dan Gieger) sent the kids off to school with upbeat music, jokes, school news, weather forecasts and even traffic reports for the parents listening in.

As the older children went off to school, Robin Blair hosted a show for preschoolers, called "Alphabet Soup". Blair, a children's music performer, taught the younger children about letters, numbers, shapes, colors and other things, read stories to the listeners with a soft, chirpy voice and played songs that stimulated both the mind and imagination. It was the kindergarten of the air.

Into the afternoon and evening hours, WWTC was now playing a wide variety of music that appealed to children; music by the likes of Raffi, the Teddy Bear Band, Barney the Dinosaur, the Sesame Street gang and the station's own Robin Blair, as well as pop artists such as James Taylor, New Kids on the Block and Maria Maldaur.

Songs heard in regular rotation on Radio AAHS included "Blanky Song" by the Splatter Sisters, "Itsy Bitsy Spider" by rock 'n' roll legend Little Richard, "Haircut" by Craig & Co. and "H2O What a Feeling" by Ariel, a character from Walt Disney's "The Little Mermaid".

The Radio AAHS air personalities played the role of an older sibling/ friend to the young listener. There was even a rotating staff of kid DJs, hired mostly from the Minnesota Children's Theater Company, who would share the broadcast booth with the afternoon disc jockey. Like any other radio station, request lines were set up, enabling listeners to call in and talk directly to the disc jockey. The station quickly acquired quite a legion of fans; early on, the station was reporting an average of 12,000 listeners, mostly but not exclusively children, calling in each month and within a year, some 20,000 kids had won prizes in station contests.

The Radio AAHS program schedule included radio plays, contests and other programs with "kid appeal", and call-in shows for parents on certain nights, in an attempt to get parents to listen to the station with their kids.

Radio AAHS even broadcast on a 24-hour schedule, playing the same cheerful music heard on the station throughout the rest of the day in the after-midnight hours, as a service to the kid who may be up late due to illness or nightmares or fighting parents. The owners realized the importance early on of ensuring that kids could tune in any time and not hear dead air or programming that did not relate to what they were used to hearing at that spot on the dial. Radio AAHS provided an island of security.

Brian Turner, a personality left over from the old WWTC, was the original Radio AAHS program director and morning host. "I got to know the owners pretty much right away", he recalls. "I knew then it was going to be huge. I felt and I knew in my heart of hearts that it was going to be big. But it was going to take time. It was not going to be an overnight sensation."

In addition to Turner, other DJs from the old 'TC included Michael Jaye, Stix Franklin and even, very briefly, Mike "Records" Ryan. An old-line AM rock jock who had been in radio for more than 20 years, Ryan considered Radio AAHS something of a low point in his career.

"It was very difficult", he remembers. "I hadn't any idea how to present it. I have children and I like children; I see little kids and I can talk to them face-to-face, one-on-one. If you can see what they're doing, you can get a lot more out of them than you can just trying to say 'how's Bobby and Susie doing today? Having a good time out there?' Well, no they're not, they're getting yelled at by mom and dad."

Adds Turner, "It's a hard thing to do when you're on the radio because you don't have the kid there to look into their eyes. If you have a kid and you're talking to him one to one, and if you begin the process in discourse of [being] condescending to them because you think you have to, you'll see right away that this is not the path to go according to their response to you. When you're on the radio, it's something you have to get over. It could be a hard learning curve because the kid could be two years old or twelve years old."

"You have to talk to kids with respect", says Michael Jaye. "You have to treat kids with the respect that they deserve. When I got on the air, I never changed my style from doing Golden Rock to doing Radio AAHS. If there's a secret to it, [it's] be yourself because if you're not, kids will see through you."

The Children's Radio Network was the brainchild of marketing executive Bill Osewalt in the early 1980s and was experimented with at a station in Jacksonville, Florida and five affiliated stations in the south-eastern United States. The original Children's Radio Network experiment was discontinued in 1986. The concept was then vigorously researched and Osewalt lured several people to invest in his grass-roots concept, including Chris Dahl of C.D. Broadcasting and marketing executive Cate Smith.

Osewalt and his associates knew that in order to start a nationwide network of children's radio stations, convincing other stations to sign on as affiliates, they would have to own their own station and make the format successful there first. Research determined that the best place to launch a radio network for children was in the Minneapolis-St. Paul market because of the large population of children, the high per-capita amount of money spent on education and the number of two-income, suburban (i.e. "yuppie") households in the area.

The Radio AAHS format slowly found success as those in charge figured out what elements would work and how the format should be defined but it took some time to accomplish this. In its first weeks on the air, it was indeed a fiasco. The music was not on individual cartridges but on long reel-to-reel tapes with about an hour's worth of music and open spots where an announcer cut in with commercials, a weather report or station identification.

It was also apparent that whoever selected the music on those tapes didn't bother listening closely to some of the songs. Included in the musical selection were the Chipmunks renditions of an old hit by the Knack called "Good Girls Don't" and "Whip It" by the early eighties punk band DEVO, both songs with rather risque lyrics.

"Everybody thinks if it's a Chipmunks song its a pretty good song and it's OK for kids. Not necessarily so!" says Michael Jaye. "Kids are very alert, very perceptive. 'Good Girls Don't' is a very sexual song. The whole thing is about sex and I'm sorry to report Radio AAHS does not promote sex. The Chipmunks doing it was really cute but what the Chipmunks were doing was selling tapes and that's fine but you don't and I don't have to play it."

There were also some staff announcers who would speak in a condescending sing-song voice. "Goochy-goo its Radio AAAAAHHHS! And coming up neeext, we're gonna play Sesame Streeeet!" one was heard saying. That announcer quickly disappeared.

Observes Brian Turner, program director in the early months, "Radio AAHS was frankly like a child. A child is born and then they grow but it takes a long time before they reach their full height and stature. I knew Radio AAHS was going to be much like the development of a kid. It was going to be birthed, it was going to need significant nursing at the beginning, we're going to have to change its soiled diapers, really, it was going to grow like a kid."

Turner had a number of ideas on how to nurture and raise that child but he had a few conflicts with founder Bill Osewalt, who jealously protected his baby.

"Bill and I got along pretty well at the start and then he didn't want me to be program director. I think there was a little too much micro-management there. If you're going to hire somebody to be a program director, let them program, and then respect the hiarchy that you put in place. You work down and up the chain like any business. Bill didn't have much respect for that so he and I had our falling out."

Michael Jaye was appointed program director six weeks into the format as Turner left to become program director at KJJO-FM (104.1), giving the Twin Cities its first commercial "alternative rock" station. Jaye worked with Osewalt and the management in making Radio AAHS more palatable for the young listener of the 1990s.

The music was finally converted to tape cartridge, with certain songs weeded out in the process and many more added, and he rescheduled and hired new air personalities, putting the best talents he could find in the best time slots.

Replacing Brian Turner on the early-morning "All-American Alarm Clock" program was Dan Gieger, who was hired from a country station in St. Cloud. His two kids were Radio AAHS fans and he had brought them down to the station for a tour when they suggested he apply for a job there. He was hired a week later.

Have Lunch with Michael Jaye

You can listen to Michael Jaye from 12:00 noon to 3:00 p.m. He brings you Lunchtime Theatre℠ and great music. Not only that, but he also interviews celebrity guests such as Bobby Smith of the Northstars and Randy Bruer of the Timberwolves. Did you ever wonder who selects the music for RADIO AAHS™ ? Here's your guy, Michael Jaye is also the Program Director. That means he makes sure the music is enjoyable for all our listeners. Thank you Michael Jaye for making sure the good stuff RADIO AAHS™ does is really good..

THANK YOU Minnesota For making Radio AAHS the #1 children's station in the world!

1280 AM
Twin Cities

Following Robin Blair's "Alphabet Soup" was Kat Connery, whom Jaye rescued from the station's overnight oblivion, and brought her to afternoons, and the evenings were hosted by Stix Franklin, who quickly became popular with the older kids in the audience.

One of the weekend jocks was Larry Wolf, who coincidentally had begun his radio career as an intern at tho old WWTC in the early eighties. He initially applied for a job as a receptionist but when Jaye saw that he had radio experience, he offered him an audition.

Contests with an educational edge became a part of AAHS, such as the "Brain Game" where a question about geography or social studies would be asked by the on-air host and kids would call in with the correct answer, and "Good Deed Days", in which listeners were encouraged to tell the station about an extraordinary thing they did for someone else. Kids who did such deeds received a Radio AAHS poster or other promotional merchandise and a tour of the studios.

Radio AAHS began to inspire some hard-core loyalty from its fans (not unlike the Golden Rock in its glory days) and praise from parents, teachers and day care providers. Day care centers across the Twin Cities were livened with the sound of AAHS playing everywhere.

Michael Jaye had qualifications as program director for Radio AAHS as both a seasoned broadcaster and a parent. In addition to having worked in both radio and television over a twenty-year career, he was the custodial father of two sons. He says that being a parent pulls some weight when one applies for a job at AAHS.

"When you become a parent, the instant the baby is born, God sends this bolt of lightning in you...all of the sudden you understand kids a lot more. Being a parent and being on the air helps because you can think about what the five-year-olds are doing."

They All Laughed...

The concept seemed absurd. Only on WWTC, the station that brought us the infamous "weather radio" format, could something as ridiculous as children's radio make it on the air, people were saying. Almost nobody thought it would work. Besides, the ratings services don't count listeners under 12 and, the theory went, what advertisers in their right mind would buy on a station that caters to snot-nosed youngsters anyway? As it turned out, there were plenty. Major advertisers such as McDonald's, Super Valu stores, Domino's pizza and Marigold foods were lined up as sponsors even before Radio AAHS took to the air.

Advertisers were beginning to realize that the pre-teen market was actually a lucrative one. In a day and age when kids earn large allowances, they were now buying more consumer goods than ever before. It also gave companies a good public image to advertise on Radio AAHS because there was a strong emphasis on education and wholesome entertainment. The advertiser who buys Radio AAHS would be perceived as a contributor to the community and associated with the concept of "family values".

Those who remained skeptical, in spite of the advertiser support, publicity and determination of the owners, received quite a shock when the winter 1990 Arbitron ratings came out. For the first time since 1985, the letters

WWTC actually appeared. Apparently more than just kids were listening because along with an overall (listeners 12 and up) rating of 0.5, the station scored a 0.9 rating among adults between ages 25 and 34. It took less than a year to do it. The founders and supporters of Radio AAHS proved that a children's radio format can actually get ratings.

According to research conducted by Chicago-based Strategic Radio Research, the station was the ninth most listened-to station overall in the Twin Cities among listeners between ages 4 and 44 and second behind country station KEEY-FM (102.1) among families with young children. Strategic Radio found WWTC has a listenership of about 57,000 children between ages 4-9 and 33,000 parents per week.

Longtime WWTC personality Dick Driscoll, who jokes that he put a "curse" on the station when he departed in 1985 (which coincided with it never showing up in the ratings thereafter), says of the success of Radio AAHS "I took my curse off that station!" He opines, "If that station was doing AAHS with Brian Short still owning it, it would still not show up in the ratings."

Rebecca Fogel, whose father had worked for WWTC long before she was a glimmer in his eye, became a big fan of Radio AAHS herself. When Arne Fogel accompanied her on a tour of the station, he recalls actually recognizing some of the equipment in the production studio from when he worked there some ten years earlier, although the station had moved three times since.

"Ugly" Del Roberts says sardonically of the Radio AAHS phenomenon, "It goes to show what a lot of the population thinks of the rest of the radio stations."

Radio AAHS and the Gulf War

On January 16, 1991, war erupted in the Persian Gulf as the United States began massive air strikes against Iraq. On Twin Cities radio, regular programming was preempted or interrupted for war coverage; but things were different at WWTC. How to handle an event like this was difficult for a radio station aimed at children.

"The war was probably the hardest decision I ever had to make as a program director", Michael Jaye told Scott Briggs of the *Twin Cities Reader*. "I couldn't go on the air and say the United States has just launched a massive attack on Iraq. I had to say that there are some things going on in the Mid-East, and you need to talk to your mom and dad about it."

The following day on the weekly call-in show "Families Today", children with questions and fears about the war in the Gulf were invited to call in and express their feelings about it as the sympathetic and reassuring adult hosts listened and explained things in an intelligent manner.

Some of the children who called in expressed passivist desires for peace and compromise. One kid asked if his house could be blown up in a missile attack. Others asked of the possibility that a relative in the service could be killed.

News anchor Coleen Needles of WCCO Television was in the Radio AAHS studio that evening and she helped explain the war to the kids in terms they could understand. She likened Saddam Hussein's takeover of Kuait to a man moving into a neighbor's house uninvited.

In the weeks that the war raged on in early 1991, Radio AAHS made a point of not talking about it. "Kids like security and we haven't changed because of the war", Jaye told the *Reader*. "I like being that island — the place kids can go when they've had enough."

Movin' On Up

It was quickly realized that WWTC's cramped studio and offices at 215 South 11th Street in downtown Minneapolis was no longer suitable for Radio AAHS. There was little space for what the owners planned on turning into the flagship for an entire network of children's radio stations and the headquarters for the company. The downtown location made a lot of parents uneasy about bringing their children in for tours of the station and the space was being rented from former WWTC owner Brian Short.

In October 1991, WWTC-Radio AAHS and it's parent company moved to new facilities in what had been the St. Louis Park branch of First Federal Savings & Loan, at 5501 Excelsior Boulevard. It was a large, two-story building, complete with a working clock tower, in a nice, relatively safe Minneapolis suburb, and C.D. owned the building.

The Radio AAHS studio and operations were located on the first floor of the building with the corporate headquarters on the second. There was now much more space for the station and the company to expand and room for the new equipment and facilities required to syndicate the Radio AAHS format via satellite to stations nation wide. There were plans of turning the old bank vault into a prize vault for contest winners but it was instead turned into a production studio and engineering area.

WWTC had finally found success as Radio AAHS. The longtime ne'er-do-well station had found its niche and was well supported by its target audience. But there were problems that still remained on the business end.

In early 1992, it was reported that the station, since being taken over by Children's Broadcasting Corporation, had suffered a net loss of $313,500 on revenues of $627,000 from its inception as Radio AAHS up to November 30, 1991.

While the station had an impressive line of advertisers, there still wasn't enough to keep it in the black. Many potential advertisers weren't convinced that the pre-teen set was that great of a consumer base.

"We are dealing with a huge audience that is hard to quantify", station owner Chris Dahl told the *Star Tribune* of Minneapolis.

Other potential advertisers assumed that Radio AAHS was a non-commercial educational station. The new broadcast facilities were the biggest financial drain yet on the station and some pessimistic financial backers questioned its ability to stay in business.

In March 1992, it was announced that Radio AAHS would offer one million shares of stock to the public at an initial price of $1 per share. The proceeds would be used to pay off cash advances granted to the station by its investors, to buy another station in another market and to build up working capital.

The offering proved successful. All million shares were sold by April 1 and within a year, they were trading at about $3 per share. This gave the station capital to finally launch the satellite and start up the long-anticipated Radio AAHS network.

Meanwhile, Bill Osewalt, one of the originators of AAHS, stepped aside as executive vice president, making way for Gary Landis, a Los Angeles radio executive who believed in the concept enough to leave the second-largest market in the country to come to Minneapolis.

With the departure of Osewalt, who owned the Radio AAHS name and other intellectual properties associated with it, the company solicited possible new names for the station and for some programs and games heard on it. But research showed that the name recognition was too valuable and that any name change would be detrimental to the success of the station. The company purchased the the Radio AAHS name and other properties from Osewalt and his associates in time to go national. The unicorn mascot, however, was eventually retired for a more hip "sound wave" logo.

Radio AAHS went national on October 25, 1992 when KIDR in Phoenix became the first independently-owned Radio AAHS affiliate. The station adapted a toll-free request line number and by January 1993, there were AAHS affiliates in Salt Lake City, Baltimore, Denver, Washington, D.C., St. Louis and Abilene, Virginia with programming originating from the studios of the Twin Cities radio station still identified at the top of the hour locally as WWTC, Minneapolis-St. Paul.

But there were the drawbacks to becoming a national network. Recalls Larry Wolf, " Any radio station, when it goes syndicated or satellite, loses any local feeling that it has. You just can't make up for it in the local weather bits that you do."

Within two years of the network's inception, Radio AAHS was heard in some forty cities, streching from KYAK in Anchorage, Alaska to WZKD in Orlando, Florida. But Radio AAHS didn't lay a golden egg everywhere it went. Sometimes it just layed an egg.

"We were surprised and upset by the lack of response", Carmen V. Nardon of WMXH Radio in Wilkes-Barre, Pennsylvania told the *New York*

Times. Having dropped AAHS in September, 1993, after only three months, the station jumped from 39th to 15th place in the market.

Fly Robin Fly

As the "child" known as Radio AAHS grew into maturity, so came the growing pains. In addition to the hassles and heartaches on the business end in trying to turn AAHS into a national sensation, the conflicts, clashing egos and politics that permeate almost every radio station plagued this one as well.

Those on the staff who didn't understand or live up to the Radio AAHS mission were weeded out early on. As time went on, there were those who left the station in frustration and anger and those who were practically escorted out the door by armed guard. Chris Dahl, while usually fair and even-tempered, also ran the place with an iron hand.

Radio AAHS tried hard to be politically correct with much on-air preaching about environmental causes, which in turn lead to an occasional complaint from an adult listener accusing "liberal bias".

The station attempted to reflect diversity by giving women and minorities prominate on-air positions. Scott Sherman, the original afternoon announcer was bumped to overnights to make room for Kat Connery, a talented young woman who had worked in small-market television and with the Renaissance Festival and was popular with the kids. She however was eventually fired and replaced by Don Michaels a former personality at KDWB-AM (630) and KQQL "KOOL 108" FM, where he had also been program director.

The station's record on diversity was called into question when former weekend personality Dorthea McKnight Ojeda, an African-American, filed a complaint with the Department of Human Rights claiming that race was the reason she was passed over for three full-time jobs at AAHS. When it was made public by *Star Tribune* gossip columnist C.J., Chris Dahl defended "All the wonderful, good things that Children's Broadcasting [Corporation] has done in communities...should [not] be overshadowed by a disgruntled employee's attitude about why they left the station."

Robin Blair, the chirpy, some would say ditzy host of the midday "Alphabet Soup" program for pre-schoolers, had the highest profile at Radio AAHS, having been established as a childrens' music performer and business owner in that field in addition to her radio duties. Her signature sign-off was "Thanks a lot for being part of the show today. You are very special." With her prominence at AAHS, however, came a highly publicized dismissal.

In the summer of 1993, while going through a "crises in my personal life", as she explained it to *Star Tribune* columnist Doug Grow, she lost a payroll check and had asked the station to issue a new one and put a stop order on the lost check. The following summer, she happened to find an old payroll check in the pocket of a suit.

As she explained to the sympathetic Grow, "I remembered back to the time and remembered how hard it was...I said to myself this is either the check they put a stop order on or it's another check that I didn't ever deposit because I was such a mess. I figured if it was the lost check, the stop order would end it and that would be that..."

Robin Blair

The check came back stamped "stale dated" and she was charged four dollars by the bank. But on Monday, June 27, 1994, after telling her young listeners "You are very special" for what would be the last time, she was called into the office and promptly fired. The next day, "Alphabet Soup" with Robin became "Avenue A" with Amy.

Blair immediately hired an attorney and contacted the press, including Grow, who likened Chris Dahl to Scrooge and a Honduran boss.

An attorney for Radio AAHS explained that the company found her explanation for the check "implausible", adding "It's not a financial issue. It's a character issue. It doesn't matter if it's about $10 or $10,000. We don't have a minimum threshold for honesty." He suggested that Blair should have told the company about the check before attempting to cash it and that the bank had initially subtracted the amount from the company account (although it was never added to Blair's).

She contacted some former employees about being witnesses in a possible lawsuit, many of whom were more than happy to have the opportunity to testify. But she wasn't a favorite among all of her co-workers.

Larry Wolf, a Radio AAHS personality who had been dismissed sometime earlier for reasons that were never clearly explained to him, recalls that Blair could sometimes be difficult to deal with.

"As we were getting ready to do the satellite, Robin was not very good at being accurate, on time, playing commercials at whatever time. [On the network] there were certain things that had to be done. You had to do a break every ten minutes.

"Robin *did not* want me to run her board. She did not want anyone in the studio. It was *her* show, *her* time. She would say things to Michael Jaye like I have bad body odor. She was always trying to get me that way and finally they just gave in to her. Robin could be a very mean, cruel person and very underhanded. But I wasn't the only one and I didn't feel the worst of her wrath."

A former office manager told the *Star Tribune's* C.J., "She was not good with the kids. She had a real fake front and was real rude to the kids on her bad days...she'd walk in and I'd be able to know by the look on on her face that it was my job to keep the kids away from her that day."

One-time program director Brian Turner observes, "Robin's always had the agenda of Robin first but you can't slight a performer for that...I like

Robin well enough but her on-air demenor I really didn't care for a whole lot. It was very talk-down, very condescending in that regard."

Sniffed Doug Grow at the end of his more sympathetic portrayal of Blair, "And AAHS keeps playing things like 'What the World Needs Now Is Love Sweet Love'."

The Great Disney Screwover

Internal struggles aside, Radio AAHS received much praise and national attention. Favorable articles were published everywhere from the business press to the *New York Times* to *Rolling Stone*. It was profiled on the network morning shows and "Entertainment Tonight", among many other places, and it was drawing the attention of national advertisers. Yet Radio AAHS continued to struggle, ending each fiscal year at a loss.

It looked like the troubles would be over, the full potential of Radio AAHS would be realilzed when in December 1995, it was announced that Children's Broadcasting Corporation had struck a deal with ABC Radio Networks for help in national sales, affiliate relations, research and marketing. Under the terms of the agreement, ABC would fund the growth of Radio AAHS and would have the option of acquiring up to twenty percent of Children's Broadcasting Corporation, which would potentially add some $11 million to the ballance sheet.

ABC had just been acquired by the Walt Disney Corporation a few months earlier, which seemed to make the deal all the more sweeter. The Disney name is synonymous with children's and family entertainment so it seemed like a marriage made in heaven. Disney gave AAHS the opportunity to broadcast from Disney theme parks and develop other tie-ins.

A stock ayaylist predicted that with the deal, revenue would jump 81 percent and losses would shrink from 88 cents per share in 1995 to three cents per share in 1997. After that, everything would be Disney World.

Eight months later, ABC/Disney abruptly severed ties with Radio AAHS. It was announced that they would be starting their own radio network for kids to compete directly with AAHS, using the Disney name. On the day of the announcement, Children's Broadcasting Corporation stock dropped 22 percent, closing at $6.12 1/2, down from $7.87 1/2 the previous day. It soon dropped into the three dollar range.

While Radio Disney premiered in the Twin Cities on the former KQRS-AM (1440) in November 1996, with programming originating from Dallas, Children's Broadcasting sued ABC Radio and Disney in U.S. District Court for fraud, misrepresentation and neglegence, among other things. The suit alleged that ABC and Disney executives visited the Minneapolis company, under the guise of fostering the alliance, "to provide Disney with the oppertunity to collect confidential information to facilitate defendants' plan to unlawfully enter the children's radio market in direct competition with Children's (Broadcasting)". The suit claimed that an executive said her job for the next six months was to learn as much as possible about the company. Disney and ABC, for their parts, denied any wrong-doings.

As the losses and legal costs for the company mounted, it was announced in June 1997 that Children's Broadcasting would sell all 13 of its owned-and-opperated stations, including WWTC, to New York-based Global Broadcasting Company for $72.5 million.

A few weeks later, it was announced that Radio AAHS would exist after 1997 only as an on-line service and that the signal broadcast on AM 1280 in the Twin Cities would be going dark.

Nancy Rosen and the author, Febuary 18, 1995.
(Authors Collection)

EPILOGUE

As WWTC sinks in the abyss, the ripple effects of its turbulent past continue to be felt.

On Saturday, February 18, 1995, former WWTC personality Nancy Rosen managed to bring together in one room an eclectic gathering of people who had been involved in one way or another with the beloved Radio 1280 at some point in their careers. Everyone from Glen "Big Daddy" Olson to Radio AAHS host Stix Franklin were there.

What had been initially planned as a reunion of Rosen's own co-workers snowballed into a "happening" that encompassed every format, era and phase the station had been through from the time the WWTC call-letters were first spoken. Air personalities, office staff, sales people and just about everybody that could be located were invited.

"Millions of Dollars Worth of Talent That Was Never Paid Accordingly" read the cover page of the guest list. Among those who signed it, "Ugly" Del Roberts, Arne Fogel, Brad Piras, former general managers Charlie Loufek and Mike Mc Anally, one-time owner Brian Short, newsman John Farrell, Magical Mystery Tour host Ray Erick, Weather Radio announcer Pamela Kay, Jean Ocken, who was an office manager there in the sixties, one-time engineer Al Flom, Alan Freed, Dave Dworkin, Scott Stevens, Gary Rawn, Dick Ervasti, Adam Abrams, Stuart Held, numerous behind-the - scenes people and of course Dick Driscoll, who probably knew more people than anyone since he had worked with different staffs, managements and owners over the years.

A collage of photos leant by the participants greeted the guests at the door, oldies music blasted from an old 'TC "Sound of Music Machine" and a wall carrying the title "In Memory Of" paid tribute to the WWTC people who most definitely couldn't be there, including Bob Short, Justin Case, Steve "Boogie" Bowman, Bob Allard and the beloved traffic coordinator Sandy Charles.

It was the quintessential love fest. Old buddies exchanged war stories, old rivals made amends and those who never met shared a common brotherhood of having been associated with that strange place known as WWTC. It wasn't always the greatest place to work, but because everyone struggled equally to keep it afloat, a unique bond formed with the troops, not unlike on the battlefield. Few work places would inspire such sentiment.

In spite of the egomania that media people are notorious for, no fist fights or bitter arguements broke out. There were a lot of handshakes, hugs and a few tears though.

The golden oldies format that the letters WWTC will probably be best remembered for remains something unique that hasn't happened again and probably never will.

Not that oldies music hasn't happened again; oldies are big business now. Practically every pop song that charted in the sixties and early seventies has been prostituted into television commercials or movie soundtracks. Today's FM oldies stations, unlike WWTC, are streamlined to the point that there are 100 or fewer tunes played in the entire format. Personalities, just like on every other FM station, are straightjacketed to the point that they are almost useless. A computer might as well be doing everything.

The overly-familiar oldies are played with such high repetition that the fond memories fade and the classic guitar riffs and lyrics become nothing more than a tiresome, bland background that nobody bothers thinking about anymore. Favorite songs have been destroyed in the name of exploiting baby boomer nostalgia to move product.

The radio industry itself has changed enormously since the days when the letters WWTC actually meant something to people. Some would argue that most of the changes have been for the worst.

While running a radio station in the Twin Cities no longer means having to struggle for survival under the shadow of WCCO, it now means going up against an entire chain of stations owned by the same corporation. Changes in federal regulations lifting ownership limits of media outlets now make it possible theoretically for one conglomerate to own every radio station, television station and newspaper in one city, controlling virtually every bit of information available to consumers. Fewer than half a dozen companies own almost all of the forty-some radio stations in the Minneapolis-St. Paul market and there will be fewer still as those companies merge. Control of information and entertainment will be further consolidated.

Proponents of media consolidation argue that there will be a better variety of programming if stations aren't competing for the same listeners. But without competition, there is no incentive to improve the product, leaving listeners with a wide variety of insipid, mediocre or downright bad radio formats. No real choice, period.

As the radio industry loses its creativity daily, it is sometimes fun to turn off the radio, and dig out the old records, cassettes and CDs. As more people program their own music, the programmers on the other side of the dial might get the message.

INDEX

SUPPORT LOCAL MUSIC

Get involved.

Contact the **Minnesota Music Movement**–
A non-profit organization that brings
Minnesota musicians and fans
together to do splendid things!

ORDER FORM

Please send _____copy(s) of FIASCO AT 1280 - THE RISE AND HARD FALL OF A TWIN CITIES RADIO STATION By Jeff Lonto.

I am enclosing $10.95 per book, plus $2.62 shipping per book. (Minnesota residents add 6.5% sales tax.)

Name ____

Address ____

City __________________________ State ______ Zip __________

Z•7

Studio Z-7 Publishing
813 Marshall St. NE
Minneapolis, MN 55413-1816